HAM Radio Operator's Guide

2nd Edition

Carl J. Bergquist KG4AIC

Other Books from PROMPT® by Carl J. Bergquist

HWS Laser Design Toolkit

Telephone Projects

Build Your Own Test Equipment

IC Projects

Video Hacker's Handbook

HAM Radio Operator's Guide

2nd Edition

Carl J. Bergquist
KG4AIC

©2001 by Sams Technical Publishing

PROMPT© Publications is an imprint of Sams Technical Publishing, 5436 W. 78th St., Indianapolis, IN 46268.

International Standard Book Number: 0-7906-1238-0
Library of Congress Catalog Card Number: 2001089474

Acquisitions Editor: Alice J. Tripp
Editor: Cricket A. Franklin
Assistant Editor: Kim Heusel
Typesetting: Cricket A. Franklin, Kim Heusel
Indexing: Cricket A. Franklin, Kim Heusel
Cover Design: Christy Pierce
Cover Photographs: Kenwood USA
Graphics Conversion: Phil Velikan
Illustrations and Other Materials: Courtesy of the Author

PRINTED IN THE UNITED STATES OF AMERICA

9 8 7 6 5 4 3 2 1

Contents

Section 3 201

Section 4 265

Introduction

So you want to become a HAM (amateur radio operator)? GOOD CHOICE!!! This field offers almost unlimited opportunity to explore both radio and electronics. It has also become relatively easy to join our ranks. Relative that is, to what it was like 20 years ago.

When I first became interested in HAM radio, there were no study guides to speak of. There were books written to help aim you in the right direction, but they looked like the New York City telephone directory. Virtually every aspect of radio that could possibly appear on your written test was covered. All of this made for some mighty tedious study.

If that was not enough, there was the dreaded CODE!!!! Oh yes, Morse code, all those "dits" and "dahs" (dots and dashes) that made little or no sense when first encountered. Personally, I feel it is a prevarication that Samuel Morse invented this stuff. I think it was some sadistic alien from the planet Zorbanon II desperate to punish the human race. Of course, that is just my opinion. Hmmmm! Anyway, you had to be able to listen to those bits of sound at a pretty rapid pace and understand what they meant. So, between those two and other priorities, I lost interest in HAM radio until just recently.

I discovered the Federal Communications Commission (FCC) had come up with the No-Code Technician class of amateur radio license. YEAH!!! That was for me! I also found out that there were books published that literally provided all the possible questions (and their correct answers) that could be asked on the written exam. Man, this business of HAM radio was looking better all the time!

Now, this is not to say that you don't have to do some studying to get your ticket (operator's license). It does take effort on your part to get one of the books, read through it, and learn what is being presented. (Or at least remember it long enough to take your examination.) However, if you possess a sincere desire to become a HAM, you should find it all fairly simple.

As stated, with the books available, a precise idea of what is on the test is available. Most of it is quite easy to follow. All of this is covered in section one.

Another very nice and convenient aspect of new HAM radio is Volunteer Examiners (VEs). These guys and gals live in your area, are at least General class HAMs, and keep you from having to travel long distances to take your test. A volunteer examiner team (VET) consists of three VEs, but a lot more is said about this in section one.

With the old system, the FCC administered the exams, and if you weren't lucky enough to live in one of the designated cities, you had to go to the closest one. Additionally, the tests were not given nearly as often as they are now. Because of the positive changes to testing and licensing, there has never been a better time to join the fascinating and delightful hobby of amateur radio.

With your Technician class license, you can explore a multitude of domains ranging from voice communications to amateur television (ATV) to satellite contacts. By the way, I traverse this aspect in great detail, in section two. Like me, I am sure you will be amazed at all there is to do out there in the world of the airwaves. Section 3 covers some of the finer points of "on-air" operation, equipment use, and propagation. This is designed to help familiarize you with amateur radio procedure and customs. To round out this text, I have included a fourth section which explores some beneficial, yet inexpensive, projects designed to dress up your new HAM shack. These have been selected for their ease of construction and pragmatic application.

On that note, let me encourage you to venture forward. The realm of the amateur radio operator is one of abundant fellowship as well as dedication to and love of the service. Sure, there are some bad seeds on the air out there, but they seldom create much of a problem. Hence, I'll lay odds you never regret your expedition into this remarkable hobby.

Most importantly HAM radio is not only radios, antennas, and frequencies. It is people -- people who come from all walks of life. You may find doctors, salespeople, lawyers, auto mechanics, truckers, Ph.D.'s, retired civilian and military employees, homemakers, writers (hmmmm...again) and a variety of others involved in this hobby.

The radios, antennas, frequencies, and all the rest become important to amateur radio but only because of the people in amateur radio. It is their commitment and adoration for the hobby that makes it what it is. This makes it well worth being part of it! Have fun and enjoy HAM radio!

~ Section 1 ~

FCC
Licensing
Scheme

There's a New Sheriff at the FCC

Over the years that amateur radio has existed, there have been a multitude of different structures set up for the process of licensing HAM operators. In the very beginning, there wasn't any governmental control at all over the process. If you had the knowledge and experience to build your own equipment, you made up a call sign for yourself and you were on the air.

However, that didn't last too long, and soon the federal government was in on the act. The Communications Act of 1934 set the first guidelines — at least for the United States — concerning wireless operation, and this was by no means limited to amateur radio. Commercial radio stations also fell into its grasp, but it did establish the first set of rules and regulations regarding HAM radio operation.

As the years went by, numerous revisions to these original rules were adopted, the number of license classes was changed, and changed again, and changed again, and the requirements were modified what seems like endless times. For example, at one time the now-defunct Novice ticket (license) was only good for one year, which meant you had to upgrade to General class within that year or lose your license altogether. At one time, the FCC conducted all testing and not always at convenient times and places. Additionally, the Morse code speed skills have been adjusted a number of times with proficiency rates ranging from five words a minute to 20 words a minute, depending on the license class.

So, when the Federal Communications Commission (FCC) announced yet another restructuring of amateur radio licensing, it wasn't taken with a great deal of surprise. On December 30, 1999, the FCC announced its long-awaited "restructuring" of HAM radio licensing to take effect April 15, 2000, and with the announcement came some significant changes to the hobby.

Probably of greatest interest to most amateur radio operators and prospective operators are two areas that experienced very noticeable change. First, prior to the December proclamation, HAM radio consisted of six license classes (Novice, Technician, Technician Plus, General, Advanced, and Extra). The proclamation reduced that to three (Technician, General, and Extra).

The second major change involved the often-controversial Morse code speed requirement. The new rate is set to an across-the-board five words per minute, which represented a significant reduction over the previous 13 words a minute for General/Advanced and 20 words a minute for Extra classes. Now, once the five-words-per-minute code test is passed, a HAM can advance to either General or Extra by taking only a written test. That is, one for the General class and one for Amateur Extra.

Furthermore, if you hold a Novice or Advanced class license, you can keep it as long as you want. Either will be renewed indefinitely as long as you don't upgrade to a higher class. However, after the April 15 effective date, no new Novice or Advanced licenses will be issued. As for the Technician Plus class, it is history. If you hold a Tech Plus ticket, it is now considered a "Technician" license with high-frequency (HF) privileges.

There were a number of other changes, but, for the most part, these don't have a major effect on new licensees or upgrades. Hence, I won't dwell on them. The entire order is available from the FCC either in hard copy or on its Web page if you would like to review it.

Naturally, with new license classes comes new question pools, but these didn't change drastically. The pools in effect right now will change in the next few years, but for the present, there are only a few new questions included with each new test. Basically, the Technician and General exams use slightly modified pools, while the Extra test combines the old Advanced and Extra pools with some new questions. The new study guides for the tests have been completed and are available from the companies listed in the source list. Just look for the testing material notation.

So, that is the "quick and dirty" look at the FCC's license restructuring of amateur radio. All in all, it makes becoming a HAM that much easier. Hopefully, both the numbers of new applicants and upgrades will increase over the next few years as a result of this action. Now, let me take you through each license class and provide information about the test and privileges for each.

Technician Class

Written Test is Element 2

As I said above, the Technician test hasn't changed all that much. It has increased from a 30- to 35-question multiple-choice test taken from the new pool of 384 questions. And, 26 correctly answered questions — 74 percent — earn a passing grade. The questions asked on the Tech exam cover FCC rules, safety, radio operation, and a basic understanding of electronics. None of this is problematic, although some of the questions can be a little tricky at times. The study guides will not only help you understand the material, but also give you the correct answer to each question. As always, the exact questions and answers are provided, but the order of the answers can be varied on the exams. This is done primarily to prevent total memorization of the test materials.

With that said, let's take a look at some sample questions taken from the new Technician pool. This will give you an idea of what to expect on the exam. As previously stated these are multiple choice in nature, and, in many cases, one or more of the answers is pretty ridiculous.

To give you an idea of what each question's number means, here is how it breaks down for the first question, T1A02. T1 indicates the question involves Commission's Rules, while T1A questions specifically regard "Basis and purpose of amateur service and definitions." Last, 02 indicates this is the second of 12 possible questions in the T1A group. This is information more for curiosity sake than anything else, as you don't have to know this to pass the test.

Okay! Here is question T1A02:

What are two of the five purposes for the amateur service?

A: To protect historical radio data and help the public understanding of radio history.

B: To help foreign countries improve communication and technical skills and encourage visits from foreign HAMs.

C: To modernize radio schematic drawings and increase the pool of electrical drafting people.

D: To increase the number of trained radio operators and electronics experts and improve international goodwill.

Mike KG4BQX is a Technician class HAM. His wife Pam works third party while studying for her ticket.

The correct answer is D. While most HAMs are concerned with protecting historical data, enjoy visits from foreign HAMs, and would have no problem with a larger pool of electrical draftsmen, these do not fall within the realm of the purpose of the amateur service. None of the ideas is bad, but the last answer best describes the service's function.

Question T1A06:

What must happen before you are allowed to operate an amateur station?

A: The FCC database must show that you have been granted an amateur license.

B: You must have written authorization from the FCC.

C: You must have written authorization from a Volunteer Examiner Coordinator.

D: You must have a copy of the FCC rules, Part 97, at your station location.

The correct answer is A. This one could get a little confusing regarding the "authorization from a Volunteer Examiner Coordinator" (VEC), as most tests are given by Volunteer Examiner (VE) teams. These teams are not coordinators, and thus, the Certificate of Successful Completion of Examination (CSCE) issued by the team when you pass a test is not "authorization" from a VEC. They merely indicate that you have passed the specified examination(s). The final authority on who is licensed and who is not is the FCC database.

Question T1A07:

What are the U.S. amateur operator licenses that a new amateur might earn?

A: Novice, Technician, General, Advanced.

B: Technician, Technician Plus, General, Advanced.

C: Novice, Technician, General, Advanced.

D: Technician, Technician with Morse code, General, Amateur Extra.

The correct answer is D. The new FCC licensing structure includes only three licenses, but the Technician class does distinguish between a straight Tech and a Tech that has passed the five-words-a-minute Morse code test. That second Tech (Morse code Tech) does have additional operating privileges on some of the high-frequency (HF) bands.

Question T1B02: (T1B covers frequency privileges authorized to the Technician control operator VHF/UHF and HF).

What are the frequency limits of the 2-meter band in ITU Region 2?

A: 145.0 – 150.5 MHz.

B: 144.0 – 148.0 MHz.

C: 144.1 – 146.5 MHz.

D: 144.0 – 146.0 MHz.

The correct answer is B. The International Telecommunications Union (ITU) has broken the world up into three regions, and the United States falls into Region 2. Since the 2-meter band in the U.S. is 144 to 148 megahertz (MHz), and this would have to be uniform throughout Region 2, the second answer is the right one.

Question T1B08:

What are the frequency limits of the 80-meter band for Technician class licensees who have passed a Morse code test?

A: 3500 – 4000 kHz.

B: 3675 – 3725 kHz.

C: 7100 – 7150 kHz.

D: 7000 – 7300 kHz.

The correct answer is B. As I stated earlier, a Technician that has passed the Morse code test does have privileges on some of the HF bands. I will cover this in more detail at the end of this section. However, here the question relates to the 80-meter band, which is 3500 to 4000 kilohertz. So, right off the bat you can eliminate the last two answers, as they are frequencies in the 40-meter band. Next, since Morse code Techs are only allowed on a portion of the 80-meter band, and answer A covers the entire band, that takes care of A, leaving only answer B. Don't be too concerned about these frequency ranges, as virtually all the study materials and other HAM-related documents (contact logs for example) have the bands broken down by frequency and privileges.

Diane N2UEC is very happy to be a Technician with HF privileges. Her husband Tom WB2ZKA helped get her started through his radio club in New York.

Question T1C01: (T1C covers Emission privileges authorized to the Technician control operator VHF/UHF and HF).

On what HF band may a Technician licensee use FM phone emissions?

A: 10 meters.

B: 15 meters.

C: 75 meters.

D: None.

The correct answer is D. Again, this information is available in most study guides, but the bottom line is that a Technician licensee only has privileges on frequencies above 50 megahertz. The Morse code Technicians have privileges in the 15-, 40-, and 80-meter bands, but only in the CW section. They also have single-sideband (SSB) phone privileges on 10 meters, but that is restricted to 28.3 – 28.5 megahertz. Furthermore, the FCC only allows frequency modulated (FM) phone in the 29.5 to 29.7 portion of 10 meters, and a Morse code Tech is not allowed to operate there.

Question T1C05:

What frequencies within the 2-meter band are reserved exclusively for CW operation?

A: 146 – 147 MHz.

B: 146.0 – 146.1 MHz.

C: 145 – 148 MHz.

D: 144.0 – 144.1 MHz.

The correct answer is D. That portion of the 2-meter band is reserved for continuous wave (CW) operation, but legally CW can be sent on the entire band. This is part of the voluntary "Band Plan" and is consistent with the HF bands where CW operation is usually found at the beginning of the band.

Question T1C08:

What emission types are Technician control operators who have passed a Morse code exam allowed to use on frequencies from 28.3 to 28.5 MHz?

A: All authorized amateur emission privileges.

B: CW and data.

C: CW and single-sideband phone.

D: Data and phone.

The correct answer is C. As stated before, Morse code Technicians are allowed to operate single-sideband (SSB) phone on the 28.3 to 28.5 section of the 10-meter band, but they are also allowed to send CW there as well.

Well, that should give you at least an adequate idea of what your Technician written test will be like. I chose these questions because they are very typical of the test as a whole and provide good examples of the pool content. Next, let's look at what privileges you will receive with your Technician license.

Technician License Operating Privileges

The process of restructuring the amateur service had no change on Technician licensees regarding frequency privileges. It allows the Tech to operate on 17 of the 27 amateur radio bands and with up to 1500 watts of power. The following is a breakdown of the more often employed bands, but operation on some additional microwave frequencies is part of the deal.

General class licensee Karen KA1BYP with her husband Lew K1AZE, an Extra class HAM.

BAND	FREQUENCY	MODE
6 Meters	50 – 54 MHz	CW
6 Meters	50.1 – 54 MHz	Phone
2 Meters	144 – 148 MHz	CW
2 Meters	144.1 – 148 MHz	Phone
1.25 Meters	222 – 225 MHz	All
0.70 Meters	420 – 450 MHz	All (70 Centimeters)
0.33 Meters	902 – 928 MHz	All (33 Centimeters)
0.23 Meters	1270 – 1300 MHz All	(23 Centimeters)

As stated above, there are a number of bands above 1300 megahertz that are open to Technician licensees, but these are used mostly for experimentation. The equipment gets a little cunning at these high frequencies, and that tends to make them less attractive to the mainstream amateur radio operator. However, if you feel comfortable up in the microwave region, a lot of interesting communications can be done there.

Morse Code Technician Operating Privileges

Code Test Is Element 1

In addition to all the above, passing the five-words-a-minute Morse code test brings operating privileges on the following band and frequencies.

BAND	FREQUENCY	MODE
80 Meters	3.675 – 3.725 MHz	CW
40 Meters	7.100 – 7.150 MHz	CW
15 Meters	21.1 – 21.2 MHz	CW
10 Meters	28.1 – 28.5 MHz	CW
10 Meters	28.3 – 28.5 MHz	Phone

These added bands give the Technician a good taste of what the high-frequency (HF) world is like. Note the emphasis is on CW operation, and this was originally done to encourage the HAM to become proficient with the code and upgrade to a higher license class. Now, of course, only the five-words-a-minute test exists, and the beauty of the restructured system is that all that is required to upgrade is passing a written test. For me, that was good news. I tried to prepare as much as possible for all tests and didn't feel at all uncomfortable taking the written exams. But, that code test! It terrified me, and I had to take it twice to pass. However, I'm certain one try will get you through.

General Class

Written Exam is Element 3

Here again, the General exam hasn't changed much. There are some new questions in the new 385-question pool, but, for the most part, the test relies heavily on the old General class pool. By the way, you may have noticed that the pools are roughly 10 times the size of the actual test. Just a little trivia! Like the Tech exam, the General class test is 35 multiple-choice questions from the pool, and 74 percent, or 26 correct answers, is a passing grade. The subject matter is a little tougher with this one, but again, the study guides will help you to understand the material, as well as give you the correct answers. As always, the questions and answers are exact, but the order in which the answers appear on the test may vary from what is seen in the study guide.

So, let me show you some examples of what you might see on your General class exam. This time, the numerical designations will begin with a "G," which stands for — you guessed it — General!

Question G1A05: (G1A covers General control operator frequency privileges)

What are the frequency limits for General class operators in the 20-meter band?

A: 14025 – 14100 kHz and 14175 – 14350 kHz.

B: 14025 – 14150 kHz and 14225 – 14350 kHz.

C: 14025 – 14125 kHz and 14200 – 14350 kHz.

D: 14025 – 14175 kHz and 14250 – 14350 kHz.

The correct answer is B. This comes from the FCC-allocated frequency plan that can be found in the study guides and other HAM-related documents. The first frequency range refers to CW operation, while the second indicates phone operation.

Question G1A10:

What class of amateur license authorizes you to operate on the frequencies 14025 – 14150 kHz and 14225 – 14350 kHz?

A: Amateur Extra class only.

B: Amateur Extra, Advanced, or General class.

C: Amateur Extra, Advanced, General, or Technician class.

D: Amateur Extra and Advanced class only.

The correct answer is B. Here the point is being made that a higher than General class license also allows operation

Phil K4OZN is an Extra class HAM licensed since 1975.

on the frequencies allocated to the lower class, or, in this case, General ticket. Since a General class HAM can operate on 14025 – 14150 and 14225 – 14350 kilohertz, then the Advanced and Extra class operators can also use those frequencies. However, the Technician class, as in answer C, cannot.

Question G1B01: (G1B covers Antenna limitations; good engineering and good amateur practice; beacon operation; restricted operation; retransmitting radio signals)

Up to what height above the ground may you install an antenna structure without needing FCC approval unless your station is in close proximity to an airport as defined in the FCC rules?

A: 50 feet.

B: 100 feet.

C: 200 feet.

D: 300 feet.

The correct answer is C. You may, without receiving approval from the Federal Communications Commission (FCC), erect an antenna up to 200 feed above ground, which makes for a really good antenna system. Rule of thumb on antennas is, "The higher the better!" While there are exceptions to the rule, it is nice to be able to reach that height without having to get FCC permission. Local regulations might be another story, however. The part about proximity to an airport involves Federal Aviation Administration (FAA) regulations regarding obstructions at the end of a runway and will only apply if you do live very close to an airport.

Question G1B02:

If the FCC rules DO NOT specifically cover a situation, how must you operate your amateur station?

A: In accordance with general licensee operator principles.

B: In accordance with good engineering and good amateur practice.

C: In accordance with practices adopted by the Institute of Electrical and Electronics Engineers.

D: In accordance with procedures set forth by the International Amateur Radio Union.

The correct answer is B. This question really requires good ol' walkin'-around sense. Remember that in the United States, the ultimate authority on radio operation is the FCC, which always requires employing good engineering and amateur practice. Groups like the IEEE and IARU, while respected and knowledgeable, have little to say about official operating rules. Hence, answer B best describes the proper conduct.

Question G1B05:

Under what limited circumstances may music be transmitted by an amateur station?

A: When it produces no dissonances or spurious emissions.

B: When it is used to jam an illegal transmission.

C: When it is transmitted on frequencies above 1215 MHz.

D: When it is an incidental part of a space shuttle retransmission.

The correct answer is D: Be advised that transmitting music over amateur radio bands is TABOO as far as the FCC is concerned. However, it appears they do cut the astronauts a little slack as long as it isn't intentional. So, unless you are an astronaut on a space shuttle mission, stay away from music!

Question G1B06:

When may an amateur station in two-way communication transmit a message in secret code in order to obscure the meaning of the communication?

A: When transmitting above 450 MHz.

B: During contests.

C: Never.

D: During a declared communications emergency.

The correct answer is C. Like transmitting music, coded messages designed to obscure the meaning of said message is TABOO! The operative word here is obscure, as some coding is allowed to control remote locations and the like, but if the purpose is to conceal the meaning of the communication, is it illegal!

Question G1C01: (G1C covers transmitter power standards; certification of external RF-power amplifiers; standards for certification of external RF-power amplifiers; HF data emission standards).

Mike KT4XL, an Extra class operator, at his station.

What is the maximum transmitting power an amateur station may use on 3690 kHz?

A: 200 watts PEP output.

B: 1000 watts PEP output.

C: 1500 watts PEP output.

D: The maximum power necessary to carry out the desired communications with a maximum of 2000 watts PEP output.

The correct answer is A. This frequency is within the "old" Novice bands where output is restricted to 200 watts peak envelope power (PEP). This rule, by the way, applies to all license classes even though no additional Novice licenses will be issued. Hence, as of this writing, even Extra class operators have to keep their power to 200 watts or less at this frequency.

Question G1C02:

What is the maximum transmitting power an amateur station may use on 7080 kHz?

A: 200 watts PEP output.

B: 1000 watts PEP output.

C: 1500 watts PEP output.

D: 2000 watts PEP output.

The correct answer is C. This frequency lies in the normal amateur radio 40-meter band, and the full legal limit of 1500 watts PEP is allowed. Incidentally, there was a time when the legal limit was 1000 watts peak envelope power.

Question G1C06:

What is the maximum transmitting power an amateur station may use on 3818 kHz?

A: 200 watts PEP output.

B: 1000 watts PEP output.

C: 1500 watts PEP output.

D: The minimum power necessary to carry out the desired communications with a maximum of 2000 watts PEP output.

The correct answer is C. One of the precepts of good amateur practice is to use the minimum amount of power to get the job done. That is, as answer D suggests, as little power as possible to achieve clear communications with another station. However, the D answer is corrupted by the maximum 2000-watt PEP output, as 1500 watts is the legal limit allowed on the HAM bands.

Question G1D05: (G1D covers: Examination element preparation; examination administration; temporary station identification).

What minimum examination elements must an applicant pass for a Technician license?

A: Element 2 only.

B: Elements 1 and 2.

C: Elements 2 and 3.

D: Elements 1, 2, and 3.

The correct answer is A. The technician ticket is earned by passing the Technician written exam only (element 2). All the other answers would earn the applicant a higher class of license.

Question G1D06:

What minimum examination elements must an applicant pass for a Technician license with Morse code credit to operate on the HF bands?

A: Element 2 only.

B: Elements 1 and 2.

C: Elements 2 and 3.

D: Elements 1,2, and 3.

The correct answer is B. In addition to element 2, to receive high-frequency (HF) privileges, the applicant must also pass the five-words-a-minute Morse code exam, or Element 1. That places the operator in the Morse code Technician category.

Here again, this random sampling should give a reasonable idea of what to expect on the General class written exam. As you probably have noted, these questions are a little more complex than the ones on the Technician test. However some diligent study from the preparation materials will get you ready. Remember, a perfect score is great, but not essential in passing the exam.

General License Operating Privileges

With your General class license comes privileges in all 27 amateur radio bands; these are the same as the old General class. In addition to everything the Technicians have above 50 megahertz, the following table shows what you gain in the high-frequency arena. The move from Technician to General represents, by far, the greatest jump in frequency access, and really opens up the exciting world of HAM radio.

BAND	FREQUENCY	MODE
160 Meters	1.800 – 2.000 MHz	All
80 Meters	3.525 – 3.750 MHz	CW
80 Meters	3.850 – 4.000 MHz	Phone (75-meter band)
40 Meters	7.025 – 7.150 MHz	CW
40 Meters	7.225 – 7.300 MHz	Phone
30 Meters	10.1 – 10.15 MHz	CW
20 Meters	14.025 – 14.15 MHz	CW

(continued)

BAND	FREQUENCY	MODE
20 Meters	14.225 – 14.350 MHz	Phone
17 Meters	18.068 – 18.11 MHz	CW
17 Meters	18.11 – 18.168 MHz	Phone
15 Meters	21.025 – 21.2 MHz	CW
15 Meters	21.3 – 21.45 MHz	Phone
12 Meters	24.89 – 24.99 MHz	CW
12 Meters	24.93 – 24.99 MHz	Phone
10 Meters	28.0 – 29.7 MHz	CW
10 Meters	28.3 – 29.7 MHz	Phone

I think this list clearly illustrates the extensive increase in frequencies the General class license provides. Nearly all the amateur radio spectrum is at the disposal of the General operator, with just an additional 150 kilohertz reserved for "Extra only" activity. In fact, only the 80-, 40-, 20-, and 15-Meter bands don't allocate the entire band to the General operator. Also, notice that the lower frequencies or beginning of each band is traditionally reserved for Morse code (CW) operation. This is part of the "band plan" existing among U.S. HAMs (and observed by some foreign countries) that sets aside certain sections of each band for specific purposes. This will be covered in greater detail later in this text, but it is worth noting at this point that the band plan is a volunteer program and, while endorsed by the FCC, is not dictated by the Commission.

Amateur Extra Class

Written Exam is Element 4

Here, we have a 50-question exam taken from a 665-question pool, where 37 correct answers (74 percent) provide the passing grade. The new pool is essentially a combination of the old Advanced and Extra pools, with some new questions added and is somewhat larger than 10 times the size of the exam itself. This makes for a little extra study, but I found most of the information easy to comprehend, especially when you examine the explanation accompanying the questions in the study guides.

At this point, you might ask, "Why bother to upgrade to Extra when I will only receive an additional 150 kilohertz of frequency spectrum?" It is a valid question. However, there are some equally valid reasons for the upgrade. For starters, I have personally found a lot of activity in the Extra class sections of the bands. As a writer, I often work some late hours (things are a lot quieter at night), and I sometimes don't get on the air until near midnight. At that hour, the General segments of the bands frequently have closed down and activity is found only in the Extra allocations. Hence, there are times when it is downright pragmatic to have your Extra ticket.

Another reason involves a part of the hobby known as "DXing" (Distant Exchange). This part I really like, and it caught my attention early on. The ability to sit in my house and talk to people all over the world still fascinates me. For me, at least, this was the principle reason I wanted to become a HAM. I have since found many other facets of amateur radio that share my attention, but being able to carry on a conversation with another amateur operator in Sydney, Australia, for example, will never cease to amaze me. With that said, again the Extra class frequency allocations come in handy. Often, much of the DXing happens in those areas of the bands. So, if you think this aspect of the hobby will appeal to you, I strongly suggest you consider getting your Extra ticket.

Another incentive for many HAMs is simply *pride*! As of this writing, the Amateur Extra class is the highest pinnacle you can reach in amateur radio. I can honestly say from my experience a vast majority of HAMs are not at all class conscious. We love the hobby and care little about the license class of the operators we communicate with. But, for some, there is a personal pride in reaching the top. This, of course, is an individual decision that each amateur radio enthusiast must make on his or her own.

Okay! I've gone on long enough about the virtues of the Extra class license! Let's take a look at some sample questions you might see on your Extra test.

As you might expect, the numerical designations for these questions will begin with an "E." You know, as in "Extra." Now, now, now! Be kind! And, the first category is "E1A." This covers Operating standards: frequency privileges for Extra class amateurs; emission standards; message forwarding; frequency sharing between ITU regions; FCC modification of station license; 30-meter band sharing; stations aboard ships or aircraft; telemetry; telecommand of an amateur station; authorized telecommand transmissions; definitions of image, pulse, and test. Wow! A lot of territory covered here, but it's not as bad as it looks!

Question E1A03:

What exclusive frequency privileges in the 40-meter band are authorized to Extra class control operators?

A: 7000 – 7025 kHz.

B: 7000 – 7050 kHz.

C: 7025 – 7050 kHz.

D: 7100 – 7150 kHz.

The correct answer is A. As was stated earlier, the Extra class license carries only a small increase in frequency privilege over the General class ticket, and this is one band when the difference is only 25 kilohertz. The study guides and many other amateur-related documents provide breakdowns of each band by license class.

Question E1A08:

How much below the mean power of the fundamental emission must any spurious emission from a station transmitter or external RF power amplifier transmitting on a frequency below 30 MHz be attenuated?

A: At least 10 dB.

B: At least 40 dB.

C: At least 50 dB.

D: At least 100 dB.

The correct answer is B. This question involves standards set forth in the FCC rules and regulations regarding transmitter emissions. The term "spurious" here refers to the part of the signal that is not intended to be transmitted (a harmonic for example). In other words, something other than the fundamental emission at the intended frequency. These extraneous parts of the signal can be filtered out, to a point, in an effort to reduce any interference to other stations, and the FCC sets specific degrees of filtering. In this case, the Commission wants a drop of at least 40 decibels in the level of the spurious emissions.

Question E1A12:

If your packet bulletin board station inadvertently forwards a communication that violates FCC rules, what is the first action you should take?

A: Discontinue forwarding the communication as soon as you become aware of it.

B: Notify the originating station that the communication does not comply with FCC rules.

C: Notify the nearest FCC Field Engineer's office.

D: Discontinue forwarding all messages.

The correct answer is A. Here, the operative word is first. Notifying the originating station of the error is a good idea, but not the first thing to do. First, get that message off the air! And believe me, the nearest FCC Field Engineer's office doesn't want to hear about this. It is not within its jurisdiction. Last, there is no need to discontinue all messages simply because one message is in violation. This question does illustrate how effective the study materials can be in educating you to the FCC rules and regulations. It is a good idea to obtain a copy of Part 97 of the FCC rules that cover amateur radio operation and to read them. But, just studying for the tests will give you a good start on understanding the FCC code.

Question E1A14:

If an amateur station is installed onboard a ship or aircraft and is separate from the main radio installation, what conditions must be met before the station is operated?

A: Its operation must be approved by the master of the ship or the pilot in command of the aircraft.

B: Its antenna must be separate from the main ship or aircraft antennas, transmitting only when the main radios are not in use.

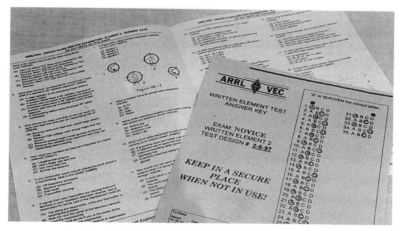

Typical example of test materials. Not scary at all, are they?

C: It must have a power supply that is completely independent of the main ship or aircraft power supply.

D: Its operator must have an FCC Marine or Aircraft endorsement on his or her amateur license.

The correct answer is A. In this scenario, the location of antennas and station power systems would be part of the approval process from the ship's master or aircraft commander. Hence, getting the station approved by the proper authority would be the appropriate action.

Question E1B02: (E1B covers: Station restrictions: restrictions on station locations; restricted operation; teacher as control operator; station antenna structures; definition and operation of remote and automatic control; control link.)

Outside of what distance from an FCC monitoring facility may an amateur station be located without concern for protecting the facility from harmful interference?

A: 1 mile.

B: 3 miles.

C: 10 miles.

D: 30 miles.

The correct answer is A. This question simply tests your knowledge of the FCC regulation regarding the minimum distance from an FCC monitoring station you are allowed to place your station. Hopefully, you won't have to worry about this one, but it is nice to know about it!

Question E1B04:

If an amateur station interferes with the reception of broad-cast stations on a well-engineered receiver, during what hours shall the amateur station NOT be operated on the interfering frequencies?

A: Daily from 8 p.m. to 10:30 p.m. local time and addition-ally from 10:30 a.m. to 1 p.m. on Sunday.

B: Daily from 6 p.m. to 12 a.m. local time and additionally from 8 a.m. to 5 p.m. on Sunday.

C: Daily for any continuous span of at least 2.5 hours and for at least five continuous hours on Sunday.

D: Daily for any continuous span of at least six hours and for at least nine continuous hours on Sunday.

The correct answer is A. This is an example of the FCC coming up with a compromise regarding use of the air-waves by various licensed stations. If the amateur station does generate harmful interference to a broadcast sta-tion, then the FCC wants the amateur station off the air during prime-time broadcasting. However, HAM equipment is designed to be very efficient at preventing such inter-ference to well-engineered receivers (40 dB rejection of spurious emission, etc.), so you probably will not run into this problem. But, if you do, you now know what the regu-lation is.

Question E1B06:

When may a paid professional teacher be the control operator of an amateur station used in the teacher's classroom?

A: Only when the teacher is not paid during periods of time when an amateur station is used.

B: Only when the classroom is in a correctional institution.

C: Only when the station is used by that teacher as a part of classroom instruction at an educational institution.

D: Only when the station is restricted to making contacts with similar stations at other educational institutions.

The correct answer is C. Whenever a professional teacher is demonstrating or using HAM radio as part of normal classroom instruction, it is legal for that teacher to get on the air during that period of instruction.

Question E1B11:

What special restrictions does the FCC impose on amateur antennas mounted on motor vehicles?

A: Such antennas may not extend more than 15 feet above the roof of the vehicle.

B: Complex antennas, such as a Yagi or quad beam, may not be installed on motor vehicles.

C: None.

D: Such antennas must comply with the recommendations of the vehicle manufacturer.

The correct answer is C. While some of the answers might sound at least logical, the FCC imposes NO restrictions on antennas mounted on vehicles. This is partly due to the fact that antennas on vehicles cannot reach very high elevations, which is one of the FCC's main concerns with fixed-station antennas. Hence, you can put any type of antennas you want on a vehicle. The primary limiting factor will be the vehicle itself.

Question E1B13:

What is meant by a remotely controlled station?

A: A station operated away from its regular home location.

B: Control of a station from a point located other than at the station transmitter.

C: A station operating under automatic control.

D: A station controlled indirectly through a control link.

The correct answer is D. The FCC does allow for stations to be operated via a control link, that is, a direct connection to the remote station that is overseen by an amateur control operator. That link can be either hardwired or wireless.

Question E1C02: (E1C covers: Reciprocal operating: definition of reciprocal operating permit; purpose of reciprocal agreement rules; alien control operator privileges; identification; application for reciprocal permit; reciprocal permit license term [Note: This includes CEPT and IARP]).

Who is eligible for an FCC authorization for alien reciprocal operation?

A: Anyone holding a valid amateur license issued by a foreign government.

B: Any non-U.S. citizen holding an amateur license issued by a foreign government with which the U.S. has a reciprocal operating arrangement.

C: Anyone holding a valid amateur license issued by a foreign government with which the U.S. has a reciprocal operating arrangement.

D: Any non-U.S. citizen holding a valid amateur or short-wave listener's license issued by a foreign government.

The correct answer is B. The key points here are "non-U.S. citizen" and "reciprocal arrangement." Answer C is almost correct except that it doesn't specify non-U.S. citizen. This only works for noncitizens who have amateur licenses in countries where we have the reciprocal agreement. That includes a large number of foreign countries, but not all of them. As for answer D, if the alien comes from a country that issues short-wave listening licenses, so be it, but that does not qualify as a reciprocal amateur radio license.

Alright! There you have a variety of examples from the Extra written examination. The fact that these questions tend to be more difficult, and a little more wordy than the General class questions has probably not escaped your attention. But, that is the nature of the FCC testing system. The higher you go, the more complex the information. Again, this test covers material from two of the previous tests and that explains the size of the exam, as well as the large question pool. However, studying for this test will reinforce your knowledge of both FCC regulations and good amateur radio operating practice.

Amateur Extra Operating Privileges

Well, what can I say? The Extra ticket is the "Big Kahuna!" When you get to this point, you now have privileges on all allocated HAM frequencies, which is to say that you can now operate anyplace in any of the 27 amateur radio bands. Just be sure to abide by the Novice power restrictions and the band plans. CONGRATULATIONS!

The following is a list of the frequencies available in the high-frequency (HF) bands.

BAND	FREQUENCY	MODE
160 Meters	1.800 – 2.000 MHz	All
80 Meters	3.500 – 4.000 MHz	CW
80 Meters	3.750 – 4.000 MHz	Phone
40 Meters	7.000 – 7.300 MHz	CW
40 Meters	7.150 – 7.300 MHz	Phone
30 Meters	10.1 – 10.15 MHz	CW
20 Meters	14.0 – 14.350 MHz	CW
20 Meters	14.150 – 14.350 MHz	Phone
17 Meters	18.068 – 18.110 MHz	CW
17 Meters	18.110 – 18.168 MHz	Phone

(continued)

BAND	FREQUENCY	MODE
15 Meters	21.0 – 21.450 MHz	CW
15 Meters	21.20 – 21.450 MHz	Phone
12 Meters	24.890 – 24.990 MHz	CW
12 Meters	24.930 – 24.990 MHz	Phone
10 Meters	28.0 – 29.70 MHz	CW
10 Meters	28.30 – 29.70 MHz	Phone

Naturally, all the "above 50 megahertz" frequencies and bands are included in the Extra class privileges. As I said, this is all there is! There ain't no more! But, you will be utterly amazed at how much activity and fun you will find on the HF bands. There is a veritable world of excitement out there, and with your Extra ticket, it is yours for the asking.

The Test!!!!!

The actual test is given by one of the volunteer examiner (VE) groups, or by the Federal Communications Commission (FCC). There are several organizations, such as the American Radio Relay League (ARRL), Central America Volunteer Examiner Coordinator (CAVEC) and the W5YI Group, that have VEs available to administer amateur radio tests on a regular basis, usually monthly. This is a big improvement over the days when you had to go to a designated FCC facility, where the tests were given only once or twice a year.

These groups are monitored by the FCC and have access to the Commission via normal communications and the Internet. The latter venue has greatly accelerated the speed at which the FCC issues licenses and call signs. In the old days, it might take several months to get your call sign. Today, amid electronic communications, it is possible to receive your call the very day you take the test. Alas, this doesn't happen too often, but a week is usually the longest you have to wait.

Additionally, there are Web sites HAMs with access to the Internet can check for their new call sign (see information list). Once it appears on the FCC page, you are legal to operate, even without the actual "piece of paper" sent by the FCC in hand. Some of the other sites will also confirm your new call sign.

As for the FCC testing you, this is most unusual. The Commission reserves the right to do so, but rarely exercises said right. Normally, it only steps in if there is some irregularity in the volunteer examining program. Often, in that scenario, the test comes in the form of a retest supervised by FCC officials. However, the VE approach is highly successful and well respected, and its results are seldom challenged. This is due largely to the integrity of the individuals involved.

Each testing group has its own facility, often a meeting room at a local organization such as a hospital, business, fraternal order, or civic group. When VEs are associated with a local radio club, the clubhouse or meeting place is frequently used, and they try to make the atmosphere as comfortable as possible.

Normally, the written tests are given first, then the code test if anyone is present for that purpose. However, that can be reversed or both tests can be given simultaneously if circumstances warrant it. The written tests have a separate scoring sheet that allows the VEs to grade the test quickly.

Code exams can be given in a number of different ways, but usually you listen to a tape-recorded message (QSO), or CD, and write down the Morse characters you hear. The test will include all letters of the alphabet, all numbers 0 to 9, and all special symbols and punctuation.

This test normally lasts about five minutes and is followed by a 10-question written exam. Passing either will bring you your "code" rating. The written exam questions will cover information sent in the "QSO." At one time, the VEs had a choice of fill-in-the-blank or multiple-choice answers for the written exam. But, the fill-in-the-blank format has been adopted as the universal standard. If you can correctly answer seven of the ten questions, you've done it ace!

Going back to the actual Morse code test, one minute of perfect copy is the minimum passing requirement. What this means is that with the present five-words-a-minute code exam, you will have to get 25 characters in a row correct. However, numbers and punctuation count as two characters each. Hence, a line with a lot of punctuation and/or digits can be very helpful in passing your exam. Of course, correctly reading an excess of 25 characters will also give you a passing grade (there I go again stating the obvious).

On a last note regarding the Morse test, it is given in the "Farnsworth" format. In this form, the individual letters or numbers are sent at around 15 to 18 words a minute, and the spacing between the words or groupings adjusts the overall speed to five words a minute. This will help you learn to listen for the sound of a character instead of trying to count "dits and dahs" and will make you a better code operator.

Now, as I said above, passing either the actual code test or the written test will result in success! So, you are really being given two chances to triumph here. I passed my code exam by getting a sufficient number of correct characters in a row (No, I'm not going to tell you how many), but came up short on the written exam. In the end, that was all it took.

All in all, the whole process takes no more than a couple of hours. That is, unless you plan to take all written exams and the code test on the same day. In that case, plan on being there for several hours, and it is entirely possible the

VEs will have to carry you to a nearby hospital following the exams. Just Kidding!!! It has been done, but I honestly wouldn't recommend trying it.

As a final word on preparing for your tests, let me address the subject of learning Morse code. You are sure to get advice from many people; some tactics will work for you and some won't. Learning code is a very personal thing. It takes a strategy that you feel comfortable with.

I have a hard time with Morse code. This might be a psychological barrier, or it might be that I'm just plain inept. Nah, it couldn't be that I'm inept, it must be psychological! Seriously, I first tried to learn code by the counting method and failed miserably.

For what it is worth, I will pass on what has worked best for me. Try to learn the individual characters by their sound and not specific "dit/dah" patterns. The later will make your copy slow and frustrating.

Also, use whatever study materials you can lay your hands on. This includes audiotapes, computer programs, code oscillators, portable code practice units, shortwave radios to monitor live code sessions, and anything else that sends or receives Morse code. The more you listen to the code, the faster you will learn it.

Last, but certainly not least, PRACTICE DAILY! You don't have to spend exorbitant amounts of time at it. A couple of 20- to 30-minute sessions a day will do the trick, but be consistent with your study. I found that every day was the best way (Uh, excuse the alliteration), and every day I missed practicing made having to go back and review what I had already learned just that must harder.

Conclusion

There you have it in a nutshell. Admittedly a small nutshell, but a nutshell nonetheless. Remember that you only have 35 of 384 possible questions on your Technician test, and it only takes 74 percent correct to pass! As for the code exam, if I can pass it, you can. Trust me on this!

A fellow HAM (KF4QOE) once related this story regarding his first testing session. When he asked the VEs how he had done, one of the examiners replied with a question. The VE asked if he knew the difference in the operating privileges between the guy who aced the test and the guy who made it by one correct answer? When my friend said "no", the examiner said, "None."

So, don't be afraid of the tests! Do your homework and be prepared, and I'm sure you will have very little trouble getting your first ticket.

~ Section 2 ~

You Have Your Ticket,

Now The Fun Begins!

Areas of
HAM Radio

Introduction

Congratulations! You made it! It was a long hard trek, but you handled it well and now it's time to consider what to do with your ticket. In this section of the book, the numerous areas of amateur radio that await you are discussed. We also look at the wide range of equipment you can obtain, many of the activities HAM radio has to offer and cover some of the etiquette, traditions, conventions and pitfalls of the hobby.

It doesn't require an elaborate station to get active on the bands. The photographs illustrate some vary simple stations (like mine) to more expansive "shacks" that have taken years to assemble. In the end, they all allow their owners to get on the air.

Add to all this a list of HAM lingo, and you are ready to join HAM radio. I am confident you will find this world as fascinating as I do. There are few hobbies I have been exposed to that offer such a vast range of activities. As one HAM (KE4LEX) put it, "It becomes a part of just about every aspect of my life." I have found that to be true. These days, I rarely go anywhere without at least my handheld trans- ceiver (HT) at my side.

So, let's begin our journey through the practical side of HAM radio. Practical in the sense of hands-on operation. In this context, practical is synonymous with entertaining (you know, FUN!!!). I can tell you that I had some pretty high expectations for amateur radio, but it indeed exceeded all my expectations. I have a feeling, call it an instinct if you will, that it will do the same for you.

Spheres of Communication

Introduction

One of the most amazing things I encountered when I first got into amateur radio was the remarkable number of individual realms within this hobby. That's a drawn-out way of saying that HAM radio has something for everybody. Well, just about everybody. I have this second cousin out in Texas that I doubt would be.....uh, that is another story. Never mind.

Anyway, if you enjoy radio communications, then this should be your bag. To name a few areas of opportunity, HAM offers phone, code, high frequency, very high frequency, ultrahigh frequency, television and teletype to choose from. These are just the beginning of a long list. Whatever field of radio you are interested in, HAM radio will be able to accommodate you.

With that said, let's take a closer look at as many of the domains of amateur radio as I could find. Some of these I have tried, and others I intend to try.

Phone
(Voice Communications)

This mode of operation applies to virtually the entire spectrum of HAM frequencies. Voice transmissions are allowed on at least a part of every band. And, with most bands, that part is substantial. So, if yakin' with folks is your thing, you will be right at home.

Even if it isn't, you are sure to relish the ability to talk to people all over the world. Along with Morse code (CW), phone communication lies at the heart and soul of amateur radio. Soon after it all began back in the early years of the 20th century, phone was there and has remained an important part of the hobby.

When sending a voice signal, you have three modes from which to operate. These are amplitude modulation (AM), frequency modulation (FM), and single sideband (SSB). Actually, FM involves two different modes that, in effect, accomplish the same goal (FM and PM), but I cover that in a minute.

Amplitude Modulation (AM)

With amplitude modulation, the intensity of the phone signal varies the height, or amplitude, of the sine wave. The number of actual cycles remains uniform throughout the transmitted signal, however. This oldest form of modulation is not used much anymore, except on the lower high frequency (HF) bands, or by AM commercial radio stations and television video.

Frequency Modulation (FM)

Frequency Modulation differs from AM in that the amplitude remains uniform, while the number of actual cycles is changed. This form of modulation is far more efficient in nature. PM, or Pulse Modulation, relies on changing the phase of the upper part of the cycle (sideband) in relation to the lower part of the cycle. In the end, though, the same result is achieved, hence PM and FM are virtually synonymous.

Single Sideband (SSB)

Single sideband is a system that removes all but one of the sidebands (upper or lower parts of the cycle) from the transmitted signal. Depending on which sideband you transmit, the signal is referred to as upper sideband (USB) or lower sideband (LSB), and defeats many of the shortcomings of AM. Today, this mode is by far the most popular among HAMs. SSB can be found on a majority of the amateur bands.

In a nutshell, that is phone communications. On any given day, I would estimate that far more phone contacts are made than code (CW) contacts. However, this is only an educated guess.

Continuous Wave (CW) or Morse Code

The second most popular, and the original method of communicating by HAM radio is CW. CW utilizes the "dit" and "dah" (formerly "dot" and "dash") patterns of Morse code to convey your message. This is accomplished by interrupting a continuous oscillator signal (wave) using a "key". The key is simply a specialized form of an "ON-OFF" (SPST) switch.

Aside from an honored HAM tradition of this being the paradigm form of communications, one advantage of CW is that your signal will be heard at greater distances.

This is due to the narrow bandwidth of the CW transmission not being lost as easily in the "hash" (noise) making this mode especially useful when trying to cut through bad conditions such as poor propagation or stormy weather.

A third method of information transfer is digital. This is the newest of the methods in use and is kind of a hybrid between phone and CW. The data is sent in code, but in most situations is transformed to an analog state so it can be conveyed with radio waves. For example, the information in your computer is easily sent over the radio with only the aid of a specialized modem. More about this is said later on.

Okay, now that you understand the basic methods of sending information over radio waves (you do, don't you?), here are some of the vehicles you can use to send said info. Among this data are many of the fascinating techniques that have given HAMs their reputation for ingenuity. Read on and you will see what I mean!

CW is the preferred method of communication for Bobby N4AU.

High Frequency (HF)

This segment of the amateur spectrum lies just off the end of the AM broadcast band and extends to about 30 megahertz (1.8 to 30 MHz). Many HAMs consider this prime operating territory. However, as seen in the first section, it does take a little extra work to acquire privileges in this region.

Due to the tendency of radio waves at these frequencies to bend with the earth's curve, great distances are achieved with even small amounts of power. A 100-watt HF rig can usually talk all over the world when the atmospheric propagation cooperates. And HAMs that like to work QRP (low power operation) have received amazing results with just a few watts of power.

Additionally, signals at these wavelengths tend to bounce off various layers of the ionosphere far better than the higher frequencies. That factor adds significantly to the range incurred on HF bands.

SSB dominates the phone conversations on these bands with AM and FM playing second and third fiddle. Here is where you are really going to see CW in action. While continuous wave can and will be heard on just about all HAM bands, it is this region that CW really calls home.

Low Very High Frequency (LVHF)

If you get a Technician license, this next band is one you are able to work. Classified as low VHF it is better known to HAMs as 6-meters (50 to 54 MHz). This band has declined in popularity in recent years, but still holds a respectable position. Propagation can be quite good when the atmospheric conditions are right, but 6-meters is more often used for local communications. FM, SSB and CW can all be heard on this band, with SSB being the most popular voice mode.

One problem with this frequency range is its proximity to the first six commercial television channels (54 to 84 MHz). TVI, or television interference (sometimes referred to by HAMs as Tennessee Valley Indians) is a particular dilemma, especially with those first few TV channels and/or dealing with inexpensive TV receivers. But, despite this difficulty, 6 meters has endured.

Very High Frequency (VHF)

In HAM land, this usually means the 2-meter band (144 to 148 MHz) and the 1.25-meter band (222 to 225 MHz). Both of these bands offer the amateur radio operator highly versatile performance with very little noise (hash). The problem with both is a lack of effective range. With a good antenna, equipment and location, both bands will travel 50 to 100 miles or so, unless there are extraordinary atmospheric conditions (a band opening). So, they are an excellent choice for local communications.

The 2-meter band is, unquestionably, the most popular amateur radio band in the country, and for that matter, probably the world. Some of this popularity has come from the No-Code Technician licenses, but even before that, the 2-meter band was highly appreciated.

The 1.25-meter band has never garnered the same respect. This has always surprised me. In a way, you have the best of two worlds with this band. It offers the carry power of VHF and the quiet conditions of UHF. Yet, 1.25-meters just hasn't caught on except in large metro areas. As a matter of fact, this band was originally 220 to 225 megahertz, but lack of use cost HAMs the first 2 megahertz.

Again, with both bands FM, SSB and CW are currently used, with the phone modes most prevalent. However, CW is often heard on the lower frequencies as "code practice" for technicians trying to upgrade their licenses.

Ultrahigh Frequency (UHF)

Although I might get some argument on this, for the HAM, UHF is the 70-centimeter and 33-centimeter bands. I know some amateur operators include the 23-centimeter band here, but for me, anything over 1 gigahertz is microwave. Then again, what do I know?

Anyway, the 70-centimeter band, especially 440 to 450 megahertz, is another very popular hangout for local communications. These frequencies are quiet and travel a long distance on a small amount of power. They are line of sight, meaning they do not bend, at least much to speak of, with the earth. Hence, you have to be able to "see" your target station.

They also do not like to bounce off the ionosphere as with HF frequencies. They like to go through it instead. Of course, this adds to the shorter range one can expect from UHF stations. However, when it comes to satellite communications, they do shine. A little power will go along way due to the inherent line-of-sight nature of an orbiting satellite.

Additionally, with the help of repeaters (repeaters are discussed in more detail a bit later), the range can be vastly improved. This, combined with the quiet signals, makes 440 FM ideal for mobile communications.

The 33-centimeter band is another that has not enjoyed a lot of support. Again, propagation at these frequencies (902 to 928 MHz) is strictly line of sight, which brings up the same problem as 70-centimeters. However, for experimentation purposes, this band is great. The noise level is almost nil, and the signal carry is magnificent relative to the power output. Thus, 33-centimeters has found a home among those HAMs interested in weak-signal and line-of-sight radio. It is also quite useful for short-range links. There is also plenty of space in this band; 26 megahertz to be exact.

As with all the others, CW, FM and SSB are present on these bands, although SSB is not as prevalent as on some of the other frequencies. FM is king, but now that I have said that, I will surely get a letter from someone telling me there is only SSB in his/her area. Always happens, never fails. But, I do like to get mail. Ha, Ha!!

Microwave

This brings us to the area above 1 gigahertz, better known as microwave. Aside from being useful to heat food, microwaves are a highly pragmatic region for certain types of communication. They are virtually dead when it comes to noise, and the line of sight propagation is all but endless. These signals will travel forever, or so it seems, on just a few watts of power.

Commercially, this range is used for telephone/television links, radar (police, weather, and tracking) and of course, the oh-so-famous microwave oven. The amateur radio operator can put them to good use as well. Many HAMs are involved in distance tests of these frequencies and have had surprising results. This has led to a new contesting arena for amateur radio.

As with the 33-centimeter band, microwaves are great for all types of links, both short-range around the house or shack and longer-range wireless hookups with repeaters. Here again, virtually all modes of operation will work.

Perhaps the most important aspect of this region is spectrum. HAMs have access to part of over 100 gigahertz of radio frequencies (that's 100 million, million!), and when you compare that to say the 500 kilohertz of the 80-meter band, or the 4 megahertz of 2-meters, that is a tremendous amount of space, easily more than enough for everybody.

As equipment becomes available (circuits for these frequencies are still in their infancy), I expect to see this area gain popularity. It will never furnish the range of 80-meters, but then 80-meters will never be this quiet. Hey, everything is a compromise.

Very Low Frequency (VLF)

This is a band strictly for the experimenter, and perhaps DXer, at heart. We are talking about 160 to 190 kilohertz. This is the region that gives new meaning to the word noise. However, you can work it without a license as long as your power is 1 watt or less and you use an antenna length 50 feet or shorter.

Many a simple circuit for 1-watt transmitters has appeared in various electronics and HAM magazines over the years, and a little library or Internet research can quickly produce something that appeals to you. Only if you are interested in trying out these frequencies though.

Receivers for this band are a little harder to come by, but some scanners and many HF rigs tune in this area. As I said before though, be prepared for the noise on this band. It is so bad most of the time that CW is the only really pragmatic approach to sending information. I've never tried VLF myself, but I have been told by some aficionados that it is fun. I'll take them at their word.

Repeaters

What is a repeater? What HAM, new or old, has not been asked that question? Well, a repeater is just what its name implies; it repeats a signal. This is done by receiving an incoming transmission on one frequency, then sending it out, or repeating it, on another frequency. This pair of frequencies is usually referred to as a channel.

The purpose of this arrangement is to extend the range of the incoming signal and also improve its quality and strength. This makes repeaters very popular with VHF and UHF operators. In fact, many amateur radio clubs sponsor repeaters in their area for use by their members, as well as the entire HAM community. These provide much better operation on the 2-meter, 70-centimeter, and sometimes 6- and 10-meter bands (also, there is limited repeater activity on 1.25 meters, 33 centimeters and 23 centimeters).

Incidentally, if you want to own and operate a repeater, have at it! That is both sanctioned and encouraged by the FCC. Many HAMs do have their own repeaters. Sometimes these private repeaters are closed. They are designed for dedicated use, but often they are open to other amateurs in the area.

There is little doubt that repeaters are heavily responsible for the widespread use of the 2-meter band and also, to some extent, the popularity of 70 centimeters. Once you get on 2 meters, you learn to appreciate the convenience of your local repeaters.

QRP or Weak-Signal Communications

This is an area of HAM radio that fascinates many amateur operators. QRP is pretty much confined to the HF bands, as these provide the needed propagation. 40, 20, and 15 meters are particular favorites, but other HF and even other areas of the HAM spectrum do get involved.

QRP stations can be very simple in nature. Usually a low-power (1 to 5 watts) transmitter, and some type of single- or double-conversion receiver. A simple antenna system, such as a half-wave dipole, added to the equation and you're off and running. "Working the world" on a few watts, or so goes the claim.

However, that claim is not as far-fetched as you might think. I have seen simple QRP stations reach thousands of miles and have heard stories of even more amazing performance. All I can say here is, give it a try, you may like it!

Alternate Power Stations

This is kind of akin to QRP as the two often go hand in hand, especially with stations designed for contesting. For example, the American Radio Relay League's (ARRL) annual Field Day event awards extra bonus point for alternative power station operation.

So what is alternative power? The answer to this lies in the definition of normal power. Normal power is electricity from a generator system as in either the power company or a gasoline-powered unit. Hence, alternative power is electricity generated by such things as the sun (solar), the wind, or the flow of water. True, in some instances a generator is still involved, but it is not powered by fossil fuels.

Naturally, the amount of electricity produced is going to be somewhat less with alternative methods. This is where QRP and alternative power find a partnership. The small-wattage QRP stations simply don't need the same amount of power as their higher output cousins.

Anyway, if you're ecologically minded, or think you would enjoy working a station with the bare minimum, alternative power might be just right for you. Even if it isn't your favorite, it is amazing what can be done with more natural methods of producing electricity. A tip: refrain from flying kites in electrical storms. Not a good idea!

Radio Teletype (RTTY)

This is an area similar to CW, but rather than "pounding brass" (sending Morse code with a telegraph key), the information is sent by typing it into an RTTY terminal or computer. This area of the hobby has a fairly good following, especially with contest lovers.

RTTY is basically a CW mode, but the speeds at which the code is sent far exceed the rate most CW operators can send or receive. Although I've heard stories about......well, never mind that. You know, stories are stories. However, most really good CW operators can understand at least part of the rapid-pace signal coming into their station.

Radio teletype generally utilizes three different protocols: Frequency shift 45-baud dot, Pactor and Amtor. Although a new phase shifting protocol, PSK-31, is now available, any of them can get you going in RTTY, with Pactor and Amtor providing error checking.

This might be just your thing. RTTY doesn't take a lot of expensive equipment, and it is a good way to send "mucho info" quickly. Also, with some of the modern software, the messages will be correct.

Packet Radio

With the advent of personal computers, it was only natural that an alliance would develop between the PC and HAM radio. Sure enough, it did. This alliance comes to us as packet radio. The term packet is derived from the method by which the information is sent out. The bits (0's or 1's) that make up computer talk are arranged in frames (groups or bits of information), and a packet is one or more frames sent out as a unit.

This approach differs from many other digital modes and has proven very efficient. When you hear packet coming over your radio it will come as bursts of sound. These packet bursts will arrive in rapid succession allowing a large amount of information to be transmitted in a short time, much the way your computer sends information over a telephone line.

How are the computer and the amateur radio connected? This is done through a specialized modulator/demodulator (modem) called a Terminal Node Controller (TNC). With the TNC placed between your radio and the computer, the digital computer signal is converted to audio (analog) signals that can be transmitted over the air.

Another computer/TNC/HAM radio station can then receive your signals and reconvert them to digital. This is then processed by the computer, and your message is displayed on the computer's monitor. Slick, ehh!!!!

Packet has enjoyed a great deal of success and popularity. Many clubs have special packet repeaters and/or factions within the club that are dedicated to packet operation. It can be very entertaining to watch a demonstration of this system. Additionally, when the phone lines are down, this is a nifty way to disseminate computer information. Yes sir, it can be a real lifesaver in a pinch.

A sidebar to packet is the Amateur Positioning Radio System, or APRS. This strategy provides a technique for determining the location of various stations via amateur radio and your computer. Again, a TNC is used to translate the two modes, digital and analog, so the radio can understand the computer and vice versa.

Additionally, the computer runs a software program that handles all the mapping of the APRS-equipped stations. As a bonus, short e-mail-style messages can be sent from the keyboard to other stations on the system. Right handy at times!

While I am not an expert on APRS (far from it), it is my understanding that each APRS active area has a repeater that is designated as the node for that area. These nodes can communicate with each other, hence extending the range of the APRS network over hundreds or even thousands of miles.

This has been just a quick overview of both packet and APRS, and both, if you are interested, beckon further exploration. I haven't had any hands-on experience with either type of packet yet, but I have had the opportunity to

Fred K8AJX at his elaborate station which is RTTY, Satellite, and packet-capable.

observe both in action. Those experiences have definitely whet my appetite for future digital operation.

Satellite Communications

Satellite activity is an area that has garnered substantial attention in the 1990s. It is partly due to an increased emphasis on amateur radio operators being able to contact each other via satellites and contact the space shuttles and the international space station (ISS), too.

All of this involves the Satellite Amateur Radio Experiment (SAREX) program that has promoted the inclusion of amateur radio equipment on many space flights. Additionally, HAMs have several dedicated satellites in orbit, These are used to make QSO's (radio contacts) with distant stations when the orbit brings the satellite within reach.

A common scenario is a 2-meter uplink and a 70-centimeter downlink. This arrangement allows a HAM to transmit a signal up to the satellite on 2 meters, then receive a reply on 70 centimeters. The satellite, itself, can be thought of as a flying repeater, and some long-distance contacts can easily be made on the VHF/UHF bands.

MIR, which died on March 22, 2001, would send signals back to earth via 2 meters. Now the ISS is doing the same thing (145.985 MHz is a common frequency). These signals can easily be picked up on a base station antenna when ISS makes its passes and may be packet bursts, voice communications, or Slow-Scan Television (SSTV). All of this can be a lot of fun to monitor, and if the cosmonauts/astronauts are working voice (phone), you might get a chance to talk with them, albeit briefly.

Satellites such as Oscar 27 and Sunsat provide long-range VHF/UHF communications to HAMs. The International Space Station (ISS) is coming along quite well and is now functional, furnishing additional opportunities to chat with astronauts and other space travelers.

As for keeping track of which orbital stations are in your area, the Internet is an excellent place to download freeware programs that track the satellites, space stations, and shuttles. The source list provides several addresses to obtain both the tracking software and programs for receiving SSTV images on your computer.

If satellite/manned mission tracking and contacts appeal to you, then you have made a good choice in HAM radio. The opportunities to monitor and even contact these "celestial bodies" is nearly unlimited as there are a bunch of them up there floating around.

Slow-Scan Television (SSTV)

Slow-scan television goes back many years in the history of HAM radio, but of late it has had a resurgence. As the previous topic notes, SSTV is used to send pictures from the space station and the shuttle missions, but that is just the proverbial tip of the iceberg.

Slow-scan is also used by thousands of HAMs to send images to each other. This can be done on virtually any amateur band, and unlike its fast-scan cousin, very little bandwidth is needed. In most instances, your home computer can both receive and transmit the photographs through the sound card.

The card is connected to the radio's microphone input for transmissions and to the speaker/earphone jack for reception. Available SSTV programs then control both the pictures for transmission and the incoming image signals. Trust me, this can be a lot of fun, and dare I say it, educational too.

Just kidding! I know if you are a HAM, or interested in becoming one, you are not afraid of education. No Sir!!! Not us! But, just in case you are, shall I say...challenge apprehensive, this stuff really is easy. As I said, it all runs through the sound card, and the software does the work. Well, most of it, anyway.

While the VHF and UHF bands are frequently used for this endeavor, the HF bands really started it and are still employed for SSTV to this day. 20 meters is a decided favorite with 14.230 megahertz leading the frequency pack. Some other opportune wavelengths are 3.845 megahertz for 80 meters, 7.171 megahertz on 40 meters, 15 meters at 21.340 megahertz and 28.680 on 10 meters. If you have an interest in SSTV, check these frequencies out.

Amateur Television (ATV)

Here is one subject that caught my attention back in the 1960s and never lost it. ATV gives the amateur radio operator the ability to run his or her own television station. You will be able to cover special events and/or club activities, run programs of community interest, show real-time weather conditions and televise discussion groups or programming aimed at entertaining just like a commercial television station. The only restrictions are no commercials and, like all HAM radio, no music.

However, ATV is restricted to the UHF and above bands. This is not an FCC thing (you know, edict) but due to a totally pragmatic reason; the large bandwidth required. Like commercial TV stations, your will command 6 megahertz of frequency spectrum. Hence, it is only those upper bands that have room for ATV. As mentioned before, 80-meters is only 500 kilohertz wide, and 2 meters has a mere 4 megahertz of space. Neither is large enough to support even a single ATV channel.

This world begins in the UHF band with 420 to 426 megahertz. 70-centimeters also offers 426 to 432 megahertz and 438 to 444 megahertz for this exercise in amateur radio excitement.

The 33-centimeter band allows two channels at 909 to 915 megahertz and 921 to 927 megahertz, while 23 centimeters has three channels (1240 to 1246, 1252 to 1258, and 1276 to 1282 megahertz). Additionally, ATV is permitted in parts of the microwave region where there is plenty of space.

I have been experimenting for awhile now with an ATV station that transmits on 438 to 444 megahertz, but have yet to get it fully operational (UHF is temperamental). Fear not, I will prevail. This unit is a kit and puts out about 10-watts. Look out NBC, KG4AIC is right on your heels. I can see the fear in the executives' faces. Well, at least in my mind anyway.

All joking aside, ATV offers amateur radio another versatile and rewarding avenue. I, for one, have always been captivated by television/video, and ATV has given me a vehicle to further explore that field. So far, even though my station isn't completely functional, I have thoroughly enjoyed the experience. Heck, I get a great picture on cable channel 60. It just doesn't go much over 30 feet. Not yet anyway!

Earth-Moon-Earth (EME)

Okay gang, here is an exotic one. Although exotic, it does have a dedicated, and sometimes aggressive, following among certain circles of the amateur radio world. Also called moon bounce, this is the process of bouncing your signal off the moon and back to earth. No, I'm not kidding! Talk about range! Thousands of miles can be spanned utilizing this process.

While theoretically just about any band could be used, EME usually makes its first appearance on the 2-meter band, around 144.1 to 144.2 megahertz. The 1.25-meter band also has spectrum allocations at 222 to 222.05 megahertz, while 70 centimeters offers 432 to 432.07 megahertz. 23 centimeters reserves 1296 to 1296.05 exclusively for EME. One of the rationales behind these frequencies is that they like to pass through the ionosphere and into outer space anyway. So, why not take advantage of that reality. They also easily pass back through the ionosphere.

The equipment used can range from a single-beam style antenna to complete stacked array systems. Naturally, the more elaborate your antenna, the better success you experience with EME. Power outputs can vary widely, but usually are in the 500- to 1500-watt range. Hey, it's a long trip to the moon and back. You know, more power to 'em, or something like that.

Actually, anyone involved in radar, either through the military or in civilian life, should be aware of this phenomena.

In the early days of radar, the screens often showed large echoes that were, at the time, unexplainable. These echoes were regarded as false echoes until someone happened to notice the radar antenna was pointed in the direction of the moon. Voila!!!

Well, for all of you adventurous types, this might just be your ticket to paradise. It might not be as well, but I won't get into that. At least, it is different, and I've been told the results can be spectacular. So, go ahead! Give it a try! Let me know how it works out.

Laser Communications

Some argue that this category doesn't really belong to HAM radio, and perhaps they are right. That is, radio waves are not involved in this type of communications. However, since much of the experimentation is conducted by HAMs, let's cut these folks some slack.

The principle is the same: modulating a carrier. In this case, the carrier is a beam of coherent light instead of a radio wave. Saying this media is line of sight would be an insult to your intelligence, so I won't say that, but it is!

As to results, they can be miraculous in terms of clarity, lack of noise and distance the signal will travel. But like HAMs interested in high gigahertz signals, laser communications is somewhat restricted to talking from one mountaintop to another. At least, that is the primary method needed to achieve any real range. Equipment here includes a fairly strong laser (5 to 25 milliwatts), some optics to further collimate the output beam and some more optics to collect and condense the beam at the receiving end. Amplifiers (audio that is) are used to both modulate the laser beam and to decode the detected beam.

I have done some work with this type of communication link, often called an open air link and the effect can be amazing. Amazing considering the obstacles the light has to contend with. But, then again, those hurdles aren't really much more severe than radio waves trying to conquer bad weather conditions.

If nothing else, this is definitely going to be a learning experience. The overall cost of the gear is not prohibitive, especially since surplus helium-neon lasers are abundant and cheap these days. The other components are readily available as well. With a little patience, you can have a station going in short order. Your only real problem then is finding another station to receive your signal.

Conclusion

This ends our tour of the different areas of HAM radio. I've tried to cover every region of the hobby, at least those regions I have heard about, but if I have missed your favorite, please let me know. I really would like to hear about it. So, with that said, let's move on to station equipment availibilty. You ought to like this part! I know I do!

Amateur Radio Equipment

Introduction

Oh boy! Oh boy! This part is neat! Even if you are a theorist at heart, you have to like talking about the hardware associated with HAM radio. There is a multitude of stuff you can buy, and I don't think I've met a HAM yet that isn't deeply involved in this aspect of the hobby.

Of course, this is as it should be. Really, it is hard to be a HAM without a radio. At least one radio, anyway! So, it shouldn't come as a surprise that amateur radio operators are engrossed in their gear.

And, WHAT gear! There is such a vast choice available for every area of HAM radio. The more commonly used bands do get special treatment by the manufacturers, but that's to be expected. However, the rest of the spectrum is also equitably represented.

While not as prevalent as grocery, discount, or drug stores, HAM equipment outlets are bountiful. In addition to several large-scale, nationwide companies, there are thousands of local merchants dealing in amateur radio gear which helps make obtaining your equipment both easy and pleasant.

Then, there are the swap meets, HAMfests, local bulletin boards, and the Internet to further augment the procurement process. All in all, it isn't difficult to obtain what you need and/or want at a reasonable price. It seems prices, at least for new gear, are going down all the time which is welcome news for any HAM.

If you are a new HAM, you might be asking, "What do I need?" Believe me, this is an excellent question. With so much to choose from, the final decision can be hard at times. It depends on what area you want to get involved in, your budget and the amount of room you have available for your shack. Then there are such factors as space for antennas and antenna restriction to consider. However, the assortment of available equipment allows HAMs to overcome most obstacles.

Oh, before I forget (I do that a lot these days), rather than trying to encompass all the various dealers and manufacturers, I felt it would be better to concentrate on the types of equipment available. That way you can assess what you are looking for as opposed to who makes it. Also, it will keep me out of trouble. Most HAMs are partial to one brand or another, and I don't want to look prejudiced toward any particular brand, as that always brings nasty mail.

So, without further ado, let's take a look at some of the marvelous stuff you can decorate you HAM shack with. In this section, I try to cover as much of the gear available as possible, but I may have missed something here and there. Though even if I did miss something, this segment reveals the magnitude of amateur radio equipment at your disposal.

Receivers

Back in the "good ol' days", separate receiver/transmitter stations were far more customary than they are today. The advent of the transceiver (combination transmitter and receiver) has caught the fancy of many HAMs for both monetary and utility reasons. That doesn't mean that receivers don't still play an important role in amateur radio.

In fact, most of the major manufacturers offer an extensive line of radio receivers. These often come as scanners and shortwave (SW) units and are useful tools in the shack. I have two scanners; one that monitors the local repeaters and another that keeps an ear on the National Oceanic and Atmospheric Agency (NOAA) weather radio frequencies. Both are well appreciated at my station.

At one time, names like Collins, Hallicrafters or Johnson were household names, at least to HAMs and, to some HAMs they still are. For most of us, these companies are just a part of amateur radio's history. They made great receivers, though. These receivers are still in demand when available on the used market. If you are looking for a really good SW radio, check out the swap meets and HAMfests. Don't expect to get these radios cheaply though. You will be disappointed.

As for what is readily available today, many excellent receivers are offered by the various manufacturers. Scanners, or radios that scan numerous frequencies looking for an active one, have captured the imagination of many DXers and thus have taken the lead in this area. They come in a variety of colors and flavors, with some being downright high tech.

Scanners also come as hand-held or base units. The hand-held units are battery operated portables that look a lot like 2-meter HTs, while the base models can be quite sophisticated. Both types usually cover multiple bands, often from 25 or 30 megahertz all the way into the microwave region (excluding cellular). Most have large memory banks for storing favorite frequencies to be scanned.

Additionally, scanning rates often can be adjusted and search functions allow the operator to explore a large block of spectrum for activity. Some models are "trunk trackers" which means they scan at such high speeds that they can keep up with the constantly changing trunking systems. These systems are becoming even more popular with police and other public service agencies.

So, for DXing and general interest receiving, scanners are great. However, most of them don't cover the shortwave bands (1.8 to 30 MHz) (SW). Hence, the trusty dealers have another category for this area of the radio spectrum. and, needless to say, a large number of radios to go with it.

Shortwave receivers come in a wide variety of sizes, shapes and capabilities. Probably the most popular version these days is the portable radios about the size of standard AM/FM broadcast receivers. These may have slide rule or digital tuning, may have a search feature, and even memory and alarm clock functions. As may be expected, the cost of the radio dictates how complex the design is going to be.

Base station models tend to be much more intricate. Larger in size, they are meant to sit on your shack's desk and utilize 120V AC household power. They almost always have digital displays for tuning, and many versions also have an analog meter to read the intensity of the incoming signals ("S" or "signal strength" Meter). Like their smaller cousins, they are multiband and usually have scan and/or search functions.

Again the price tag is going to be very indicative of how much you get in the radio. How much in terms of functions and features, that is. Unlike the Collins and Hallicrafter radios, which were continuous tune systems (no breaks in the available frequencies), the newer receivers tend not to cover the amateur radio bands. Mostly, they stick to the commercial SW frequencies.

Hence, if you want to monitor HAMs, you may have to add a second HF rig to your arsenal. The manufacturers are right there to help you out. Virtually all the major dealers carry a line of radios designed for just that purpose. Like the others, they come in a wide range of prices and capabilities.

So, other than some very exotic receivers that seem to come and go with time, that about covers receivers. Well, in a brief way at least. There is a lot more to this, and for that matter to all the subjects covered here, than space allows me to explore. Naturally, a little research into your area of interest is highly advised, and goes a long way toward making you a happy consumer.

Transmitters

Today, individual transmitters, except in the areas of QRP and experimental, are all but unheard of. That is, for the commercial new equipment market. Again, there was a time when transmitters were a standard part of any HAM shack, but that time was before the transceiver. Some people feel this concept has taken something away from HAM radio, but on the other side of that coin, transceivers are smaller, more reliable and, by comparison, less expensive than the original two-unit scenario. If you are interested in setting up a station that uses the separate transmitter/receiver approach, your best bet is a medium to large HAMfest. Here, vintage equipment should be in abundance, which allows you to procure what you want and need. Always, though, when buying used gear, expect the seller to demonstrate that the equipment works properly. If the dealer refuses to do that, look elsewhere.

However, as I mentioned, QRP (small power) stations almost always employ the separate transmitter/receiver strategy. So, not all is lost. Equipment designed for QRP is offered by a number of companies, and may be in kit or completed form. As a gesture of kindness, compassion and general good will, I have included a QRP station, yes, complete with individual transmitter and receiver, in the last section of this text. You'll love it! Trust me!

Sadly, the old "boat anchor" (large, cumbersome) transmitters have bitten the dust. Well, not really. You've got to watch me, I lie a lot. No! Just kidding! They still exist on the used market and can make for a very interesting HAM station. One day though, even those will be gone, and that will be dismal. However, remember that in the transmit mode, your transceiver is a transmitter, So, HAM radio does live on!

Transceivers

Ah yes, the "POX" on HAM radio according to the separate-unit purists. It is undeniable that transceivers have changed the look and feel of amateur radio. But, many regard that change as for the better. After all, exploiting solid-state technology, today's transceivers are far more reliable, much smaller and less costly than their predecessors. That should be good news to all HAMs.

Anyway, setting that controversy aside, let's take a look at what is available to you in this area. Again, many great radios are awaiting your attention on the used market. Swap meets, HAMfests and other HAMs are all good sources for this equipment. You'll notice I refer to the transceiver as a radio. In the world of the HAM, all communications gear is thought of as a radio.

For the HF operator, a multitude of systems are out there, and they range in price (new) from around $1,000 to as much as $5,000. It's another example of "more bang for your buck." These units are full-function, multimode (AM, FM, SSB, CW) powerhouses that generally start at 200 watts maximum output. Linear amps can boost that power, but that's another section.

The capabilities of the radios will utterly amaze you (unless you're comatose and/or brain dead). Through high selectivity, super sensitivity and advanced filter designs, these wonders easily work the world. I found a majority of the HAMs I've talked to feel their first HF purchase was one of the most exciting of their amateur radio experiences.

As for the remaining sectors of the hobby, every band I can think of has an extensive line of transceivers. This includes 6 meters, 2 meters, 1.25 meters, 70 centimeters, 33 and 23 centimeters and even into the gigahertz domain. Naturally, 2 meters garners the most attention, but 70 centimeters is not far behind.

Additionally, most companies provide a line of dual-band, and even tri-band, radios. These are very convenient in that you can monitor and operate on more than one band from a single radio. Don't forget that all HTs are transceivers. Many of them are also multiband in design. Can you imagine trying to carry a separate transmitter and receiver around like you do an HT?

Other features, especially in the VHF and UHF units, include crossband repeat (the ability to receive on one band and repeat that signal on another), built-in duplexing (for multiband systems) and a wide range of memory and other functions. All in all, you get a lot of radio for your money.

Remotely located moblie transceivers are becoming more and more popular all the time. These units allow you to locate the main radio under a seat or in the trunk and operate the radio via a cable-fed control module. The module can be the front panel of the transceiver or, with one company, a combination microphone, speaker, and LCD display/function control unit.

With the removable front panel variety, you are able to mount the panel in a location of convenience, such as the dash. This allows for quick removal for either operational or anti-theft reasons. With the combo mike, speaker, display models, the hand-held control unit can be removed or hidden under a seat for similar reasons.

Hence, undoubtedly your first purchase in HAM radio will be some sort of transceiver. This is not a bad thing; I don't care what some people say. Even if you do long for the "good ol' days", you will truly enjoy the transceivers.

Linear Amplifiers

Here we have an area that falls into the category of controversial. It has been my experience, limited as it is, that there are two factions within the amateur radio community regarding this subject. That is a windy way of saying some love these things and some hate them. The arguments go something like this. The FCC mandates that you use only the power necessary to complete the QSO, and linear amps increase said power way above what is necessary. That, of course, is the viewpoint of those who are against these amplifiers. The pro linear amplifiers faction maintains that the extra power is a courtesy measure, as it is impolite to force distance stations to strain to hear your signal. Also, the extra power is necessary for very long-range communications.

I'm not going to take a side on this one. Call me a coward if you want, but I've seen this argument get quite nasty at times, so I feel it is best to stay clear of it (bruck, bruck, bruck, bruck).

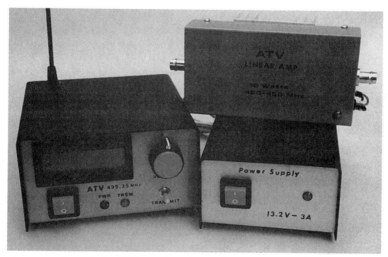

An experimental ATV station featuring a 10-watt linear amp.

Anyway, again setting the debate to one side, linear amplifiers are used by many amateur radio operators to increase their transmitting signal strength. I do use one in my automobile to boost my HT, but the HT only puts out 5 watts max to start with.

Both the new and old markets are full of amplifiers ready to serve your needs. They do come at a price though. One unit designed to produce 1K (1000 watts) will set you back about $5,000. Some others are not that expensive, and a little judicious shopping is definitely in order if and when you decide to purchase a linear.

Most of the amps use vacuum tubes in their finals (last stage) to acquire the high wattage power. Many require a high-amperage power supply to provide the needed juice. Additionally, you have to install antennas and feed-line capable of handling the large power outputs. That, of course, adds additional coins to the grand total.

If you want the increased capability, then those cost factors are probably justifiable. Once the amplifier system is installed and working, there is no question you will have greater versatility in terms of transmission range. How much depends on the band you're working. Remember, the FCC does allow up to 1500 watts on many of the amateur frequencies.

As a final comment on this subject, it is pretty much a personal decision. Both sides do have a reasonable argument supporting their stand, so let your individual needs, and maybe conscience, guide in this area.

Antennas and Associated Friends

Speaking of antennas — weren't we speaking of antennas — they are the backbone of any HAM radio system.

The better your antenna, the better your performance, regardless of your power output. Take my word on this one, there is no substitute for a good antenna — none whatsoever. I like to use a photography analogy here. You can buy an expensive camera with a motor drive and all the other whistles and bells, but if you put a $10 lens on it, you're gonna get $10 pictures. Your antenna is the same. That $5,000 HF rig is going to perform poorly on a cheap antenna system.

That isn't to say that you have to have the very best. Not at all! Many middle-of-the-road (cost wise) antennas do a superb job. However, it is best to discuss any purchase choice you feel ready to make with people who know what they are doing. Luckily for you, and for me, HAM radio is full of those people. HAMs that have been in the hobby for 10, 20, 40 years are going to be of immense help when it comes time to choose an antenna. Most of them fit the "been there, done that" category, so they know of what they speak. Request and heed their advice!

Directional Antennas

Generally, antennas are recognized as falling into one of two types; directional and omnidirectional. The directional systems are just as their name implies. They send the signal out in more of one direction than any other (there is always some radiation in all directions). Frequently called beams (for beaming the signal), they come in a variety of styles.

The most common beam is probably the Yagi. Named after its Japanese inventor, Yagis consist of a center support called the boom with cross elements spaced at calculated dis-

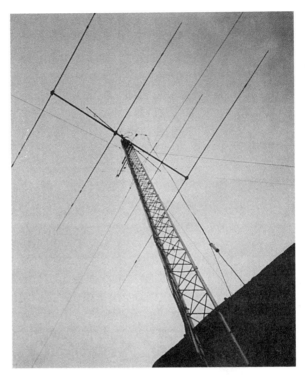

A classic 3-element Yagi beam atop a tower.

tances along the boom. The cross elements are of different lengths and can be as few as three to as many as 33 or more. The individual dimensions of a Yagi system are dictated by the frequency it operates on.

The first, or longest, cross element is called the reflector (REF) with the next element down the boom the driven element (DE). From there on down the boom, all other cross elements are called directors (D1, D2, etc.). As with much of HAM radio, the more elaborate the system, the better the performance. Yagis are no exception to this rule.

The second most common beam is the cubical quad, or just quad. These are often just two elements, but can be more than that if desired. As with the Yagi, the more elements you have in your quad, the better it is going to work. With the two-element arrangement, the first one is the driven element, with the second being a director. The feed line is connected to the driven element.

Dipole Antennas

A simple, but highly effective antenna is the dipole. Here, two lengths of wire (elements) are extended in opposite directions with the feed line connecting at the middle. For example, the center conductor of a coaxial feed would connect to one wire and the shield to the other.

The length of the dipole elements is determined by the band of choice. A simple formula for this is 234/the operating frequency in megahertz. Thus, a 10-meter dipole, for Tech Plus operation, would be 234/28.30 megahertz, or each wire would be 8.27 feet. This is for a 1/4 wave dipole, and if you want a 1/2 wave dipole, double 234 to 468 in the formula. The frequency of 28.30 megahertz is derived from the space allocated to a Tech Plus licensee. Since that space is 28.10 to 28.50 megahertz, 28.30 falls dead center. Cutting the dipole elements to the corresponding length should give you good performance throughout the designated spectrum.

Dipoles are considered to be slightly directional off the sides of the elements, as opposed to the element ends. Directionality is by no means as distinct as with a beam, though. While jumpers can be employed to connect different element sections to make the antenna multiband, dipoles seem to work best when they are resonant (cut to length) for

*A classic HF "antenna farm" featuring a 3-element Yagi.
Notice the dipoles extending out from the tower.*

one particular band. However, the simplicity of their construction makes them easy and cost effective to build. Even
if you have to build one for each band you want to work.

A slightly different take on the dipole is the inverted V where
the shape of this dipole resembles an upside down V. These
antennas are considered to be almost omnidirectional and
are often more convenient to install as the ends of the
elements extend down towards the ground. There they are
easier to reach and work with. A caution though, try to
keep the angle of the inverted V at 120 degrees or more.
Below 120 degrees, antenna performance is affected.

Helix, Helical, or Helical Phase Design Antennas

Another highly directional antenna is called the helix, helical or helical phase design, depending on who you talk with. Used mostly for VHF and above, these antennas are best described as a single element wound around and around a center element. When pointed in a given direction, the signal is amplified tremendously in said direction. Helical antennas have found a home with the weak-signal folks who really like to stretch the range of the 900 and 1,200 megahertz bands. They are great for extending the reach of weak-signal video links.

Omnidirectional Antennas

Alright, so much for directional antennas. Let's look at the ever popular omnidirectional antennas. Of all the styles available to HAMs, as well as commercial communications, the ground plane has to be one of, if not the most celebrated design. Fitting in the category of vertical antennas, when it comes to omnidirectional operation the ground plane, or some variation of it, wins the prize.

Basic Ground Plane Antennas

If you look closely at the various vertical antennas advertised in catalogs, magazines, etc., you can see evidence of a basic ground plane design. This is because the concept is so sound in terms of efficiency. The basic system does not display any gain (the term for this is unity gain), but the more elaborate designs can and do provide gain factors of 3 dB, 5 dB, and higher.

Hence, the ground plane is a HAM radio luminary. When it comes to local 2-meter communications, you can't beat a ground plane of some sort. Even a simple unity gain antenna will amaze you with that type of operation.

Discone Antennas

A second omnidirectional style is called the discone. Some argue that this is just a specialized ground plane, but I won't get into that. Basically, a discone has a center vertical element surrounded by either a cone-shaped solid element or a series of individual elements. The side elements or the cone extend downward at approximately a 45-degree angle.

The dimensions vary according to different manufacturers, but the basic design always follows the above description. Discones rarely display anything but unity gain, but like the ground plane, their performance is remarkable considering the omnidirectional pattern they emit.

So, when you're considering an omni base antenna, one of these two choices should provide you with a very workable scenario, especially when it comes to local area communications. There are more ground planes to choose from, but the discones are gaining on them all the time (excuse the pun).

Mobile Antennas

Up to now, we have discussed only base station antennas. However, there is a whole different world out there where the mobile antennas live. Designers have their hands full when it comes to mobile antennas. For starters, you don't have an established ground as you do with base station arrangements. Hence, the vehicle's body has to become that ground.

It goes without saying that the vast majority of mobile antennas are omnidirectional in nature. That is the only practical method for most situations. I mean, it is hard to drive your car and adjust a beam at the same time. You don't want to end up pushing up daisies over something stupid like that.

Be that as it may, antenna designers have come up with some mighty astonishing results within this arena. Depending on your requirements and the amount of money you want to spend, your choices range from the simple cut-to-length whip to base or center-loaded high-gain systems.

Some are shaped like cellular telephone antennas, some have magnetic mounts and others have radiator elements like a ground plane. Dealing with HF operation is where the real challenge lies. Since antenna lengths for those frequencies are substantially longer than VHF and UHF, designers have to find ways to condense that length into something suitable for mobile applications. While at the same time, try to retain performance.

A number of schemes have been applied here. Heavily loading a whip is probably the basis for most designs, and some even employ remote controlled motors that change the length of the antenna. The selections are abundant. Hence, when it comes time to equip your vehicle with a rig, you shouldn't have any trouble finding an antenna to fit your needs.

Last, but certainly not least, come antennas for our HTs (handy talkies). Often, an HT is the first radio many HAMs buy and use, but they do, by nature, suffer from restricted range. This is due to the fact that a hand held has to have an antenna suited to its size. The average HT is going to be from 3½ to 6 inches tall, and it wouldn't be very pragmatic to have a 3-foot antenna on it. Fear not, the proverbial designers have tackled this trial as well.

Their first goal is to keep the antenna as short as possible. Naturally, like so many things in electronics, a compromise has to be reached. Your particular needs will help dictate which of the many available antennas you put on your HT.

Said antennas range from the telescoping whips that extend in length to the heli-flex or rubber ducky versions that are heavily loaded fixed-size antennas. Overall, you will get better range with the telescoping variety, but the flexible rubber antennas have come a long way in recent years.

Other designs include very short, stubby types for shorter range needs, base-loaded heli-flexes, mid-loaded heli-flexes, and telescopers. Each has its benefits and drawbacks, but all are remarkable in their performance, considering their size. Again, your individual requirements mandate the type you chose.

Other Antenna Accessories

The last order of business for this section is to cover some of the many associated items on the market -- associated to antennas, that is. Your antenna is just one portion of your overall system, and it needs some accessories to complete the whole picture.

For example, if you have a base station with a beam, you need a rotor system to move the beam. This piece of equipment comes with a control unit that tells you where the beam is pointed, and the two are connected with, you guessed it, a rotor cable. That, of course, is in addition to the feed line that hooks the antenna to your radio.

Feed line comes in its own share of sizes, shapes and colors. When it comes time to buy feed line, you are going to encounter names like coax, twin lead, ladder line, and so forth. In many cases, coaxial or coax is your best choice. I say in many cases because certain antenna designs compel the use of something other, like ladder line, for example.

Getting back to coax, it comes in a variety of different sizes, designs and qualities. Usually, the larger sizes, such as RG-8 or 9913, provide the least amount of signal, or line loss. If your station is going to require long cable runs (50 feet or more), these larger size cables are a good choice.

The smaller sizes like RG-58 or 8258 are acceptable for short runs and mobile applications. The line loss is higher, but the short cable length compensates for that factor. Again, there are as many opinions on this subject as there are types of cable. For newcomers it is best to consult a HAM who has had experience with the various feed lines.

Cable connectors are another consideration with your antenna system. Here, you can chose from UHF or PL-259, SO-239, BNC, N, and several others. Naturally, they come in both male and female styles. You need to use the type that corresponds to the connector on the back of your radio(s). For example, most 2-meter radios will have an SO-239 on the back of the radio. That will need a PL-259 on the cable to make the connection.

BNC connectors are usually associated with HTs and UHF equipment, while N units are UHF and very low loss applications. Also, the type of construction materials can make

a difference. PL-259s can be purchased with either phe-
nolic or Teflon (diallayl phthalate) insulation material (di-
electric), with the Teflon considered the best. They are also
plated with nickel or silver, with silver being preferred.

So, there are a lot of factors to consider even when buying
cable connectors. This makes the hobby both fun and mul-
tifaceted. I like having choices and reasons for those choices.
HAM radio always offers an abundance of both.

Antenna Analyzers

This item could fit in the test equipment category, but since
I'm on antennas, let me hit it here. There are a number of
analyzers on the market, and like most other areas, these
come in different sizes, capabilities and price ranges. Most
all have some form of display (either an analog meter or
LCD readout) that tells you what the device is doing.

As the price goes up, so does the sophistication of the analyzer, ranging from simple measurement such as power
output and/or SWR to alphanumeric displays that visually
indicate true impedance and resonance, as well as SWR
and power. Additionally, some have a built-in frequency
counter that exhibits the operating wavelength of the input
signal. One very elaborate unit even has both analog meters
and the digital display.

Most of the analyzers have fairly wide bandwidth coverage
(1.7 to 200, 400 or 500 megahertz) as opposed to the SWR/
power bridges that are generally restricted to the HF bands,
VHF and/or UHF bands and so forth. However, if you plan
on adding an analyzer to your arsenal, it is best to check on
what frequencies it covers.

Many HAMs I have talked to that have purchased an analyzer feel it was an excellent investment. If you want to get into homebrew antennas, it is probably an indispensable investment.

Towers

Something else you may want to ponder, regarding your antenna system, is a tower. If you plan to have your antenna at heights above 30 or 40 feet, you need a tower. Especially if it's HF beam and/or vertical.

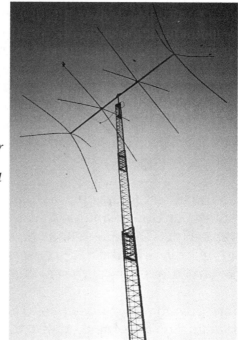

An example of a tower topped by a large 4-element, cubical quad antenna.

Now, towers are not cheap! They will put you back more than a little pocket change, particularly if you buy new. Used towers can be a good deal, but it is important to inspect them carefully before buying. Check for heavy rust, areas in the tubing that retain water and any lose or broken cross pieces. If these conditions exist, pass up that so-called good deal.

Towers come as individual sections (often 10 foot lengths) that are stacked on top of each other, styles that tilt over for easy access to the antenna, and crank-up models that literally crank up at the turn of a handle. Some are free-standing (no other support necessary), while others require guy lines, or wires coming off the tower at angles and secured at the ground to reinforce and stabilize the tower.

The gauge, or weight you need depends entirely on the size and type of antenna placed on top. For large HF beams, a sturdy heavy gauge tower is required not only to support the weight of the antenna, but also the rotor that turns the antenna. Additionally, larger antennas have higher wind resistances which cause strain on the whole system. Hence, you want a rugged tower.

If, on the other hand, you plan to put a small VHF beam for 2-meters, or UHF beam for ATV, then a much lighter duty tower will do. These antennas have neither the weight or wind resistance of their HF cousins, so a smaller, less durable structure easily supports them.

Before buying new it does pay to check out what is locally available. For one thing, these towers are cumbersome and heavy, and the freight costs can be substantial. Good sources are HAMfests, local shopper-type publications, and talking to local HAMs who may have, or know of someone who does have, used towers/tower sections for sale. It may take contacting more than one person to get everything you need, but if you follow the above guidelines, you'll end up with a darn decent tower for a fraction of the retail price.

Lightning Protection

There is an accepted principle regarding antenna systems that the higher you can get the antenna, the better your signal is going to transmit. For the most part, this is true. However, tall antennas also make great lightning rods. In fact, all antennas make great lightning rods. Unfortunately, if lightning decides to use your antenna for such a purpose — remember, lightning wants to get to the ground taking the easiest possible path — and your radio is connected to your antenna, you can kiss that radio good-bye! I'm not just whistling Dixie here! A direct lightning hit, and even a close hit, will burn that precious radio to a crisp (think bacon)!

While there is no real substitute for a nice long expanse of air (disconnecting the antenna connector from the radio), the next best thing is to install some form of lightning protection. This often comes in the form of lightning arrestors (heavy-duty surge protectors), which send a heavy surge to ground rather than through your radio.

These devices do their job in a couple of different ways, but the end result is the same. Some use a spark gap that is only activated when the voltage/amperage reaches a certain level. Some of the early homebrew lightning arrestors actually used automotive spark plugs where the threaded end was set in a copper pipe that had been driven into the ground. The other end was spliced into the feed line center conductor. In this fashion, the power was only diverted to ground when it reached a dangerous level (such as in a bolt of lightning). I'm told you can make your own arrestors this way, but I'm not sure I would trust them. Besides, there is no telling what that arrangement does to your line loss and/or SWR.

Wiely KE4LTT displays the remnants of a fiberglass antenna after it took a direct lightning hit.

Another popular design makes use of a glass envelope, gas-filled ionization chamber. In this strategy, only a damaging power level ionizes the gas, letting said power pass to ground. Both methods are commercially supported, and I have satisfactorily used both types. One additional advantage of these devices is that they help bleed off static charge build-up on the feed line.

An extra way to protect your radio from electrical storms is to use power line surge protectors. These units divert large surges directly to ground, hopefully protecting your radio gear. Again, in the end, that wide air space between your radio's power plug and the wall socket is by far the best protection. The moral to this story is that when there is lightning in your area, unplug antenna(s) and the 120V AC lines from your equipment.

Incidentally, this is excellent advice concerning your computer, as well. And, don't forget the telephone line that puts you on the Internet!

Dummy Loads

Let me briefly discuss the dummy load. This is not what it might sound like, but a substitute antenna. You DO NOT want to operate your transmitter without an antenna, as the lack of a load can do some downright nasty things to it. On the other hand, there are times, like when you are tuning up your HF rig, that you don't want the signal to travel very far.

The solution here is the dummy load. It both suppresses the signal radiation and provides the proper load to protect your equipment. They are very handy to have around the shack. They are also rated according to power. It is necessary to have a dummy load that handles the power your radio puts out.

Antennas have been called "a science all their own", "black magic" and some other names it wouldn't be polite to mention here. One thing is for sure. They can and do make or break your station's performance. Some good advice is to get to know antennas and how to use them.

Duplexers and Antenna Switches

As a sidebar to our discussion of antennas, let me touch briefly on duplexers and antenna switches. Both of these devices allow HAMs to use different radios with a single multiband antenna (duplexers) or use a single radio with several different antennas (switches). There isn't a whole lot of mystery to all this, it's simply two methods of interfacing your radio with your antenna.

The duplexer is a device that senses the frequency the radio is transmitting and allows only that signal to pass through. That signal goes to an antenna that is designed to handle more than one band, and the duplexer makes sure the signal goes only to the antenna and not into the back of the other radio. Feeding a strong RF signal directly into the antenna connection of any radio is a surefire recipe for disaster. For example, I have separate 2-meter and 70-centimeter transceivers, but I use a dual-band vertical antenna for both rigs. The duplexer makes sure that only the signal from the radio I'm working gets to the antenna. It also prevents that signal from backtracking into the radio not in use.

Antenna switches are just what you might expect them to be: a switch that selects a particular antenna and feeds it to a single radio. They are useful with HF rigs where you might want a vertical for 10-meters, but a beam or dipole with 40-meters. Using an antenna switch, you can chose whichever antenna is needed for the job.

Due to their simplicity, there isn't a whole lot to say about them. I would recommend a good brand, as some of the cheapies cause trouble. Be sure to get one rated for the radio you use. If your HF rig has a maximum 100-watt output, select a switch that handles at least double that much power. Otherwise, you run the risk of blowing out the switch.

Antenna Tuners

Rounding out the dialogue on antennas, let me mention antenna tuners. Tuners are made to match nonresonant antennas to the band you are operating on. These are most handy in the HF region, but are seen at higher frequencies as well. By installing the tuner between the radio and the antenna, adjustments can be made with variable inductors and/or capacitors that result in the needed match.

This matching not only improves performance, but also guards against possible damage to the radio. This is a risk that always exists when a transmitter is not in resonance with its antenna. As for receiving, the danger of damage is not present, but the better the match, generally, the better the reception.

Antenna tuners come in a variety of style and sizes. Some are very basic units with only control knobs. Others incorporate a SWR/power meter for a visual representation of the matching. As with everything, the more complex, the more expensive. When considering a tuner, it is best to evaluate your needs and let that be the benchmark for what you purchase. However, down the line you may require a more sophisticated tuner, so keep that in mind as well.

Handy Talkie (HT)

As I stated earlier, many a HAM starts out with a handy talkie (HT) as their first radio. This is often a good choice, as it can be kept with you most of the time allowing you to become familiar with the hobby. Like everything else we have discussed though, they come in an almost confusing number of different models.

The first considerations are what band(s) you want to work and how much money you want to spend. The bands available in HT are from 6-meters to 23-centimeters (you don't see many HF HTs), and for a new radio, the money can range, from about $150 to $500 or more.

Anne KG4MTI finds HTs useful for local communications on 2 meters and 70 centimeters.

The most popular HTs are on the 2-meter band, and many of these units are dual-band meaning they also receive and transmit on another band as well (usually 70-centimeters). 1.25-meters also has a limited inventory of HTs, and I have seen a few on 10-meters, however they were older, discontinued models similar to CB walkie-talkies.

The latest trend, as of this writing, is the tri- and quad-banders. Some companies add the 6-meter band to 2 meters/ 70 centimeters and another has included the 23-centimeter band with the normal duo. One unit has all four. This can definitely add versatility to your hand held operation.

Depending on the number of coins you want to give away, the functions and features of HTs range from basic units to hundreds of memory channels, computer cloning capability, and even access ports for such things as SSTV. One radio even has a thermometer built in. In fact, it is astounding just how much stuff they can put in such small packages these days.

With all that said, you should most definitely consider an HT for your first or, at least, one of your first radios. Beyond what we have covered, just the convenience of having personal communications with you so much of the time makes it worth the investment. Also, as you get into more and more amateur radio activities, the HT becomes almost a necessity.

Transverters

Back when I became interested in electronics (many moons ago), these were called converters, but today they go by the handle transverters. These are devices that convert the operational frequency of a radio from one band to another. For example, a 2-meter transceiver can be transverted to work in the 6-meter band. This is a handy and cost effective way to expand your HAM shack's capabilities.

Several companies handle transverters, and while I personally have never used one, HAMs I've talked to say they do a fine job. Maybe a little loss in overall performance, but not enough that you would notice.

In general, they save some money, but I kind of doubt they will take the place of a dedicated frequency rig. However, as has been said before, in electronics there is always compromise.

Meters

Here again, this subject could have been part of the test equipment section, but since meters are a major component in many shacks, I felt they deserved their own segment. Then again, maybe I'm overreacting.

Anyway, the most important meter to most HAMs is the SWR/Power variety. This gem allows monitoring of both the Standing Wave Ratio (SWR) and the transmitter power output with a single device. Usually, you have to switch back and forth between the measurements, but many manufacturers are combining two separate meters into a dual pointer (cross needle) single analog movement that keeps tabs on both measurements at once.

A quick word about standing wave ratio (more is said later). This is important! Knowing the SWR on your transmitter/ antenna system can save you both heartache and money.

SWR is measured by a ratio (1.5:1, for example), and when the ratio is too high (anything above 3:1), it can destroy your precious transceiver. Hence, you assuredly want to keep an eye on that condition.

Briefly, standing wave is the power that does not make it out of the antenna. Thus, it remains standing on the feed line. Unfortunately, this standing energy eventually backs up into your radio's final section, overloads it and well....you don't want to know! It an ugly sight! POW, BANG, BOOM and lots of smoke! Trust me, you want to save yourself and your family from this.

All kidding aside, it can be a killer in terms of your equipment and that is why HAMs are so fond of their SWR meters. You will be too. SWR is easily adjusted at the antenna. Knowing that your ratio is within tolerance saves you bolting awake in the middle of the night screaming.

The power meter simply tells you how much power (wattage) you are delivering to the feed line. While this figure is not a danger to the health of your system, well, at least not as dangerous as bad SWR, it is, at times, important to know. It can also be indicative of possible trouble with a radio. If your rig is supposed to put out 50-watts and you're only getting 35, there may be problems.

Another popular meter is the watt meter. This is similar to the power meter, except that it gives you a true reading of your power output. The power meters are close, but if you really need to know exactly what the wattage is, this is your boy! They tend to be expensive, though. Because of this, many HAMs rely on their power meters for a "close-enough-for-government-work" reading.

Next on our list is the field-strength meter. This dandy device tells you, without any direct hookup to your system, the relative strength of your signal. It can also be used to measure the in-shack RF radiation level. While it is very nice to know that your signal is getting out, there is a more significant reason to have this device. The following explains.

Knowing the field strength of your signal can be quite important, especially if your station's output is 100-watts or more. In that scenario, the FCC requires you to monitor your radiated power to make sure it falls within its safety guideline. This involves such things as maximum permissible exposure (MPE) and milliwatts per square centimeter of radiated power.

RF safety is covered in more detail in a later section, but for the moment, let me just say that RF can be a health hazard if not properly monitored and managed. For that reason, this meter is a valuable addition to your equipment list. Also, as stated, the FCC does have guidelines regarding this, and it also checks from time to time to see if those guidelines are being observed.

That covers three of the most important peripheral meters you see in HAM shacks. Much of the equipment we have yet to discuss includes some of these meters as built-in features. This is especially true of the SWR/Power monitor.

Power Supplies

Many amateur operators like to use mobile rigs in a base station capacity. They are cheaper than dedicated base station models and smaller when space is a consideration.

This is a good way to get on the air, but in this configuration you need a power supply. How big a power supply depends on what type of equipment you use.

For instance, if you intend to run a standard 50-watt mobile 2-meter or dual -band rig, it probably pulls 9- to 11-amps when in the high-power (50-watt) setting. Therefore, you need a supply that provides at least one amp more than the rating for the radio. In this case, a 12-amp supply does nicely.

A 100-watt HF rig, however, pulls in the 18- to 20-amp range when at maximum power, so a 25-amp power supply is in order. You want to be sure enough amperage is available, as too little causes problems like intermittent operation and/or possible damage to both the radio and supply.

Some advice given to me by a number of HAMs when I first got started was to purchase a 25- to 35-amp power supply, the logic was that later down the line I would add an HF rig, and it is always best to buy a supply that you can grow into as opposed to one that you may grow out of and have to replace. Very good advice indeed!

The variety of power supplies is almost as great as the variety of radios themselves. Some are linear supplies where the power is obtained from a heavy-duty voltage regulator varying load resistance (kind of a brute-force approach). Others are switching where a separate oscillator circuit varies the ON/OFF time of the system to provide the regulation. Both types are reliable methods of furnishing the needed power and each has its own fan club.

Some are of the black box design with little more than a power switch and binding posts for the output, while others have variable voltage, amperage and voltage meters, and various output receptacles (cigarette lighter, binding post, etc.). All should, however, provide a 13.8 volt output. Most amateur equipment works at 12 volts, but normally not up to full potential. Which one is right for you is your decision, a choice made according to your requirements. Naturally, the more whistles and bells the more expensive.

Another factor to be concerned about is the output hygiene. That is to say you want a clean output, free of transient signals and other noise. These maladies are usually the result of poor filtering and can be quite annoying. They can also, in some rather unique cases, cause equipment damage. So, be sure to check the specifications and/or ask the dealer about this when you're shopping for a supply. Normally, the name brands have this area well in hand.

In review, you need a supply that provides an excess of current (amperage), provides the proper voltage (usually 13.8 volts) and has a clean output. The linear units are heavier, as they have larger transformers, but that is sometimes an advantage. Whichever way you go, just follow the above guidelines and you should end up with a very satisfactory power supply.

CW Keys

As important as continuous wave (CW) is to amateur radio, it is only natural that, over the years, a number of different sending key designs emerged. The original straight key was very effective, but it was exhausting to use. So, the iambic paddle versions soon became a part of CW. The latter are, for the most part, favored by a majority of HAMs, but there is still a dedicated group that uses the straight key.

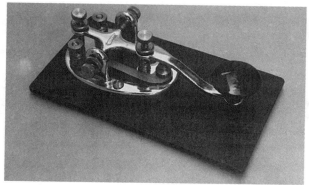

The classic straight key.

The primary difference in the paddle and straight keys is the way in which the Morse code is tapped out. With straight keys, a downward motion is applied to the key knob, whereas with the paddle keys a squeezing, or slight brushing of vertically positioned levers does the trick. The downward motion takes more effort and thus fatigues the operator faster, or so many CW operators contend.

Personally, I haven't yet gotten into CW enough to make a judgment along these lines. But, I did do my homework in terms of discussing this subject with as many CW-active HAMs as I could find. Generally speaking, the iambic or paddle keys are favored, but that is a decision that has to be based solely on personal preference. I imagine I will quickly find a favorite, as I'm sure you will, once I really dive into CW.

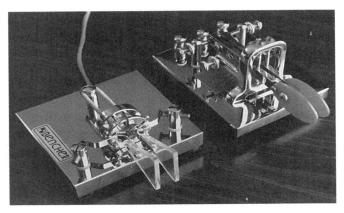

Two examples of iambic paddle keys. On the right is a Vivroplex; on the left a Bencher.

Electronic Keyers

Before I became a HAM, I always wondered what these electronic keyers were all about. I reasoned that if you were good at sending and receiving Morse code, then why would you want an electronic gadget to do it for you? Unfortunately, I had missed the point regarding electronic keyers. Or should I say I misunderstood what they were for?

Electronic keyers don't send and receive code for you, they merely smooth out your sending to make it easier to understand at the other end. No matter how well you send code, you are going to make minor mistakes in terms of intervals between characters, speed of characters and the like. The electronic keyer's job is to correct those mistakes thus making the code more understandable. Remember, code is often received under less than perfect conditions and anything to help its intelligibility is welcomed by most amateur radio operators.

As to their physical appearance, they are usually small to medium boxes with control knobs and/or switches. The simplest models may only set code speed and possible dit/dah duration, while the more intricate and expensive designs incorporate message memories, speed control, volume, tone, sidetone control, auto, semiauto and manual character creation and so forth. Depending on how much CW work you plan on doing, any or all of the features available may well be in order for the keyer you buy.

Microphones

Here is an area of HAM radio that has a rich history. Microphones are used to send voice (phone) signals (DUH!), but there is a broad spectrum of ideas concerning what is the perfect microphone. One thing is for sure: there are a whole heap of choices.

Microphones come in dynamic, electret and crystal versions with each one having its own personality. The old carbon element types are just about defunct these days. Microphones are made for both desktop base station use as well as mobile application and can be small or large.

The base station varieties range from the classic D-104, with its sleek pedestal connecting the element housing to the base (they come in black, silver, and gold), to goosenecked small-element housings sitting on control-panel-style bases. Standard dynamic studio microphones can be at-tached to adjustable arms that provide high position versa-tility. It can also be part of a headset arrangement that positions the microphone when worn.

Mobile microphones can be found in the standard hand-held, push-to-talk (PTT) variety, as lapel units with VOX (sound switching) PTT that allows the driver to keep both hands on the wheel, or as part of a headset unit. There is also the microphone/ speaker combination that clips to a shirt collar. These are popular with HT users.

Companies have hit the market with amplified types, styles with echo and other special audio effects, and units that literally blast you onto the air (usually a bad idea by the way), all in an effort to get your attention and money.

Over the years that HAM radio has evolved, microphones of a multitude of shapes, sizes, designs, capabilities, etc. have come and gone, while some have come and endured. Microphone manufacturers are quick to make use of the latest technology to improve their product, and this has led to many of the small, but mighty, microhones we see in use today.

So, if you're of a mind to replace that factory-issue model that came with your 2-meter rig, you shouldn't have any trouble finding something you like. A lack of variety is the last thing to be encountered on that shopping trip.

Batteries and Chargers

This section applies to the hand-helds (HT) and other portable radios. Most handy talkies utilize some form of rechargeable battery power source for convenience sake.

A few employ only dry-cell-style batteries (heavy-duty cells, alkaline, etc.), and most HTs offer such a pack as an accessory. There are times when dry cells save the day, such as when the power is out and you can't re-charge your rechargeable batteries.

But, for the most part, there are three basic types of bat-teries used by HTs. These are the Nickel-Cadmium (NiCd), Nickel Metal Hydride (NiMH) and the Lithium-Ion batteries. Each has its own pluses and minuses.

Nickel Cadmiums have been around the longest, and are in the process of being phased out by the other two. This is due to their less-than-great longevity (even when properly handled, four years is about the effective life of a NiCd cell), and their aspiration to develop a memory. This last one is usually the result of improper recharging and causes the cell to take on less than a full charge. It is an annoying feature that has lost the NiCd popularity over the years it has been available.

Nickel Metal Hydrides are newer than NiCds and have elimi-nated the memory problem. They can be charged over and over again without the danger of a less-than-full charge. They also display far better longevity than NiCds, which makes them a hit with HT owners. However, they are usu-ally more expensive than NiCds.

Lithium-Ions are the new kids on the block and their cost reflects that. As of this writing they are ridiculously priced, but I'm sure that will change over time. They have an even better longevity than the others, a higher capacity, and tend to be substantially smaller. Thus, once their cost comes into line, Lithium-Ions will be the preferred battery for portable application.

The other side of this coin is the device to put the charge on the battery. With any of these batteries, charging is a process of applying power of the proper voltage and amperage for the needed time. This, in effect, reverses the discharge process.

Chargers come in several different varieties. Most radios come with one of those wall-wart type plug-in gadgets with a long cord that plugs into the radio or battery. These work, but as a rule are SLOW! A better approach is the rapid charger.

The rapid charger has circuitry that increases the amperage applied to the battery during the initial charge, then backs it off for the final charging stage. This procedure allows the battery to take the charge at a much faster rate. Lithium-Ion batteries, on the other hand, already take the charge at a faster rate, so rapid charging is not as much of an issue.

That provides a respectable overview of rechargeable batteries. I'm certain more types will come along in the future, and 10 years from now, we may not even be talking about Lithium-Ions, let alone the other two. For the moment, those are the choices, so we need to make the best of them.

Terminal Node Controller (TNC)

If your plans for amateur radio include packet radio or the amateur positioning radio system (APRS), then this is a device you need. The Terminal Node Controller, or TNC, is a specialized modulator-demodulator modem/controller combination that allows your radio and computer to talk to each other.

This is done by changing the digital information from the computer into analog signals that can be sent by radio. In the reverse process, the radio signals are changed to a digital format so the computer understands them. All of this is essential for the two areas of the hobby I mentioned above.

TNCs are usually medium-sized boxes with a bunch of pretty lights (LEDs) and a switch or two. The lights keep you posted on what the TNC is doing, and the switch is normally a power ON/OFF control. Once hooked to the computer and radio, operation is automatic.

Features and functions of the average TNC are: VHF and HF operation, multi-protocol compatible, baud rates from 300 to 9600 (depending on price), and a variety of accessories to customize your TNC system. Of course, the price you shell out for the unit greatly affects the features you get, but you should be used to that by now. In the words of Yogi Berra, "Deja vu all over again."

Incidentally, as a side note, your computer can become a good friend in your HAM shack. Many other areas of this hobby can be enhanced by the use of a computer, and some outright require one. I'm speaking of domains like Slow-Scan Television SSTV, satellite tracking and HAM-related Web sites and information, to name a few. So, keep the computer in mind when setting up your shack.

Phone Patches

With the advent of cellular telephones, phone patches have lost some of their sparkle. But, before cellular, they were a big hit and still are with amateur radio clubs that sponsor and operate repeaters. As the name would imply, this device is used to patch a radio into the telephone system. Thus, you can use them to make telephone calls from your HT or mobile rig.

The area I live in has two phone patch-capable repeaters on 2-meters, and I can tell you, they come in very handy. I carry my HT just about everywhere I go, and with the phone patches, I don't have to worry about carrying the cell phone. Besides, cellular service in my area leaves a lot to be desired.

Physically, phone patches resemble TNCs and some electronic keyers. They are long slim boxes with LEDs and switches. Some are full-duplex which means they operate much like a standard landline (telephone), while others require the PTT switch to be depressed in order to talk to the other party. This blanks the receiver so you can't hear until the PTT switch is unkeyed. These are most associated with repeater systems, but can actually be connected to any two-way radio. So, if you want to put one on your base station, have at it.

That's about all I can say about phone patches. In reality, I could say more, but I think you get the idea! I mean, I have just so much space for this book so let phone patches go at that.

Filters

Virtually all radio frequency (RF) signals contain harmonics. What are harmonics? Well, keep your shirt on, and I'll tell you. Harmonics are not a musical instrument you play with your mouth. They are spurious signals that accompany your transmitted signal. They are at multiples of the actual frequency you intend to transmit. For example, if you are transmitting at 100 megahertz, you are also sending harmonics at 200, 300, 400 megahertz and on up the spectrum.

These harmonics can be quite strong, which can cause severe interference with other HAMs and radio services. It is nice to get rid of as many of these demons as possible, thus keeping you in good graces with your neighbors and fellow HAMs.

This is one way filters come into play. They handle the task of removing harmonics from a radiated signal. High- and low-pass filters are especially good at reducing television interference (TVI) which has been known to tag along with HF signals. It is highly annoying to anyone trying to watch television in peace.

These filters can be installed at the offended receiver (TV set for instance), or in line with the transmitter and antenna. In either case, pass filters do a competent job of removing most, if not all, the TVI. While a good many high/low-pass filters are designed to eliminate the TVI problem, they also can be constructed to solve similar trouble in other frequency ranges.

Another service filters perform is to clean up incoming signals to your receivers. HAMs probably put this assistance to work more often than filtering harmonics from radiated signals. When working certain bands, such as the HF frequencies, noise can be a tremendous problem. It can make hearing weaker stations almost impossible unless you can filter out that noise. See!!! There it is — FILTER.

To this end, most HF radio manufacturers carry an extensive line of accessory filters used to reduce background and other on-air noise. These are dedicated filters for specific modes (AM, SSB, CW) and some typical center frequencies would include 250 Hz, 500 Hz, 1.8 kHz, 2.4 kHz, 6 kHz, 455 kHz and 9.0106 MHz. All, of course, are designed to make working certain sections of the HF spectrum easier.

If intermod (spurious or distance interference) is a problem, you should probably look into a bandpass filter. This type is fashioned to allow a narrow frequency range to pass, while excluding all others. These are most useful on the VHF and UHF bands where a band opening (extraordinary propagation) allows distance signals to travel great distances thus causing unintentional interference.

Digital Signal Processing (DSP)

This is kind of the granddaddy of all filters in that it handles most problems with a single unit. Digital signal processing (DSP) reduces or eliminates heterodynes (those eerie wailing background noises familiar to DXers), noise and interference on nearly all modes. That includes the likes of RTTY, SSTV, FAX, packet, etc.

These units, similar in size to small antenna tuners or packet controllers, incorporate a whole bundle of different, and sometimes exotic, filters to really clean up incoming signals. Some claim to so completely reduce HF noise with SSB that you will think you are listening to local 2-meter FM. I can't confirm or disclaim that declaration, but I have seen DSP in action, and it was impressive.

Again, you can purchase DSP systems in varying degrees of complexity. The price reflects the functions and features present on any given unit. Some DSP filters offer almost automatic operation, and the high-end versions usually include a 2- to 5-watt low noise audio amplifier to boost the volume.

Preamplifiers

In the world of amateur radio, preamplifiers (preamps) are usually seen in one of two categories; antenna and microphone. Each can be an important addition to a station when the need arises.

Antenna preamps are employed for, you guessed it, incoming signal boosting. They are used more with the VHF and above bands than on the lower frequency HF range. If you plan on getting into satellite communications, I strongly advise you investigate preamps available for the 2-meter and 70-centimeter bands. Many satellites downlink with less than 1-watt of power, and a preamp can really make the difference between a successful and unsuccessful contact.

Preamps are used, however, just for the incoming signal. Hence, most of them have built-in methods to take them out of the line when transmitting. This is often accomplished with RF relays or other RF sensing and switching circuits.

As for microphones, preamplifiers are used to either boost the audio signal to the radio, or match the impedance of the audio input to the radio's impedance. In the boost mode, they can increase signal carry through saturated modulation, but it is wise not to overdo the project.

Overmodulation is not only an annoying nuisance, but also very poor operational conduct.

When employed to match impedance, the preamp can be a very handy device. Most microphones come in either low impedance (500 to 600 ohms) or high impedance (usually around 50,000 ohms). Radio audio inputs, however, may not always match the impedance of the microphone you are trying to use. In this scenario, the preamp adjusts the microphone's impedance to correspond with the radio. It's a nifty trick that can save you money over the long run.

Test Equipment

This subject can encompass a multitude of different gadgets, but for the most part, just a few will be necessary for most HAM shacks. The long list might include a voltmeter, ammeter, digital multimeter (DMM), analog multimeter, vacuum tube volt meter (VTVM), signal generator, oscilloscope, frequency counter, SWR meter, watt meter, power meter, capacitance meter and this list can go on and on.

As stated, for amateur radio purposes, only a few of these are required often enough to justify their cost. Among the requisites would be two we have already discussed, the SWR/power meter and the watt meter (marginal). To that list I would add the DMM, an analog multimeter and a frequency counter. Oscilloscopes are nice to have around, but EXPENSIVE!

So, let's talk about the multimeters and frequency counter. First off, both the digital and analog versions do basically the same job, but they do it in a slightly different manner.

For this reason, I do recommend you get one of each. They aren't all that costly anymore.

Digital multimeters measure values such as voltage, amperage, resistance, and sometimes capacitance and limited frequency (normally the audio range). Most of them also include a diode tester or transistor tester, and some have an audible continuity tester. That last one can be especially handy. What yours has depends on how much you want to spend.

Many of the analog meters have much of the same stuff, with the exception of the capacitance and frequency capabilities. There is a difference in how they take the measurement. With digital meters, you see what just happened, while analog devices tell you what is going on right now.

This may seem like nit-picking, but it really can be important in certain circumstances. For example, when tuning a transmitter circuit, you really want to know what is happening at that moment, not what just happened. Thus, a good choice for that task would be the analog meter. Again I say, "Get one of each!" Both types are readily available from just about any electronics supplier. At some point down the line, you will be glad you did. Trust Me!

Frequency counters are specialized event counters that use a timing window or gate to control the length of time the counter functions. Got that? Good! Okay, let me elaborate. An event counter continues to count pulses as long as it is enabled. To get an accurate reading on a frequency, you have to limit the enable time to something standard like one second. Since frequencies are expressed in one second intervals (20 megahertz means

20,000,000 pulse or cycles in one second), confining the count to a one-second window/gate results in a precise measurement of the input frequency.

Most frequency counters give you a choice of gate timing to expand the resolution of the counter. This is because they usually have eight-digit displays and run out of digits as the frequency increases. By using 1/10 of a second as the timing window, you increase the readable frequency from 100,000,000 to 1,000,000,000. To illustrate this, 200 megahertz would not fit on a eight-digit display unless you increase the resolution from cycles per second to 10 cycles per second. This is done by speeding up the timing gate to 1/10 of a second instead of one second. In that fashion, the first displayed digit now represents 10 hertz instead of 1 hertz. Not quite as accurate, but close enough.

I hope that explains how the frequency counter works. As to how it might be of use to you, think of the possibilities. You are able to check the frequency of various stages of a circuit when troubleshooting, and it gives a precise frequency measurement of your signal. Many counters have a built-in detector that acts as an RF receiver; hence, no direct connection to the radio is needed.

I have used frequency counters for many years, and I wouldn't be without one. These gems have saved me so much time and trouble. I definitely advocate this piece of equipment for your shack. Fortunately, frequency counters have dropped drastically in price over what they were even 10 years ago. Hopefully, you will be able to add one to your bench.

Repeaters

I have talked about repeaters before in this book, but this is an opportune time to cover them in more detail. As a reminder, repeaters are stations with the sole purpose of receiving a signal on one frequency then sending that signal out, repeating it, on another frequency. The purpose of this action is to extend the range and improve the quality of signals coming into the repeater.

This is especially helpful with mobile stations, as their signals are rarely as strong as base stations. Hence, repeaters give them that extra boost to be more efficient. However, base stations, too, use repeaters both to talk with each other and to mobiles. That further extends the capabilities of some bands.

Speaking of bands, repeaters first show up on 10-meters. From there you find them on 6-meters, 2-meters, 1.25-meters, 70-centimeters, 33-centimeters and 23-centimeters. There are also a number of ATV repeaters around on 70-, 33- and 23-centimeters. If you don't already have one, I strongly recommend the American Radio Relay League's (ARRL) repeater directory which provides a complete listing of all U.S. repeaters. They can be purchased at any HAMfest or directly from the ARRL. This listing is real handy if you travel a lot or when on vacation.

As can be seen, repeaters are used on the upper bands where propagation does not provide long-range signal carry. The HF bands, with their ability to curve with the earth's surface, carry well enough by themselves, and repeaters are usually not necessary.

Repeaters, or machines as they are sometimes referred to, are often owned and operated by amateur radio clubs. This is as a service to the community where the clubs are located and to the HAM community in general. Remember, amateur radio is a service. However, any licensed HAM can own a repeater and many do.

Many of these privately owned machines are open, or available, for general amateur use. Some, though, are closed repeaters that are reserved for a certain group or individual. These normally require a special tone code to access them. I have heard arguments that these should not be allowed. The reasoning here is that no one owns a frequency, and closed machines are depriving other HAMs of that frequency. This dispute has apparently gone on for some time, and both sides endorse what they feel is a legitimate case.

To wrap up our discussion on repeaters, let me correct a slight misnomer I made at the outset. I stated that a repeater's ONLY job was to repeat a signal. Actually, that is the repeater's primary job, as they are also used to provide telephone system access through auto-patches. This still falls into the area of repeating a signal, though.

Also, for your knowledge, the two frequencies utilized by a repeater are called a frequency pair, and a good number of repeaters use access tone codes that can be set on most radio equipment. These are subaudible tone (although they can be heard) control signals that are transmitted each time you press the PTT switch.

They range in frequency from 67 hertz to 250.3 hertz and are used to open the repeater. For the most part, this tone access is installed to prevent interference from another repeater using the same frequency pair. While repeater coordinators try to keep machines using the same pair far enough apart as not to interfere with each other, it doesn't always work that way, particularly during extraordinary propagation or band openings. The access tones normally take care of the problem.

By the way, repeater coordinators are volunteers that assign frequency pairs by region. This is to keep systems with identical pairs at great enough distances so they do not interfere with each other.

If you work 2-meters and/or 70-centimeters, you will soon learn to love your local repeaters. They can and do make communications on those bands far more reliable.

Miscellaneous

This is the section where we mop up those stragglers wandering around out there. Actually, I couldn't possibly get all the stragglers. Some of those little devils are downright slippery. But, I do hit the high points regarding some additional stuff you might be interested in and/or want to include in your setup.

Speakers

Let me start with speakers. Many HAMs like to hook external speakers to their rigs for better audio. I have two mobile radios being used as base stations. They work great, but the internal speakers on both are located facing toward the bottom of the case. This makes sense for mobile use, as the radios are usually mounted under the dash, and the sound can then bounce up off the floor of the vehicle. But, for base station application, this directs the audio down into the bench.

The answer to this dilemma was to use external speakers. See, I'm always here to help you with these complicated problems. Alright, alright! Be nice! With the appropriate plugs, I simply connected two small units that came off a miniature boombox radio. They do a great job.

If you want to get fancy, many dedicated speaker assemblies are offered by various manufacturers. They range from just a speaker in a cabinet to amplified units with adjustable volume. Many of the HF transceivers offer accessory speakers in cases that match the radio's style, and some even have power supplies built in for convenience.

There are a bunch of external speaker possibilities out there. Another good source is computer companies. They offer full lines of both amplified and nonamplified speakers for sound card operation. These normally have excellent sound quality, and on the surplus market, can be had for a song. Well, not quite, but darn near. Look around you! You may be surprised at what you already have on hand.

Power Distribution Centers

My next victim, uh.....subject, is distribution centers. Power distribution, that is. I have thoughtfully included, in the last section, plans for such a system, courtesy of Mike KT4XL. This is an item you will want once you start adding radios to your shack. Most power supplies offer only one pair of binding posts for access to their output, and that one pair can get crowded when you start trying to hook up several pairs of wires.

If that was not bad enough, wire has this annoying habit of wanting to squeeze out from under those binding posts as you try to tighten them. Not only could this cause a poor connection, it could cause a short. A short could damage or destroy your power supply. Not a nice thought, is it? That beautiful, and expensive, 35-amp power supply kaput over something as silly as too many wires on the binding posts.

The answer is, of course, a power distribution center. I mean, why else would I be talking about them? Take a quick look at Mike's center, project #3, and you can see that they are very easy to construct. If that's not your bag, several companies have units already built and tested for you to choose from.

Common elements to all these centers are several sets of binding posts or banana jacks hooked in parallel and wire/terminals heavy enough to handle high amperage (this varies by model, usually starting in the 10-amp range and extending to 50- or 60-amps).

With the input leads connected to the power supply's output, you are able to attach your radios and other equipment to the distribution center. Now your mind can be free of the horrors of exploding equipment due to a power short.

Mike's center also includes an analog voltmeter to keep tabs on the voltage level. A current meter can also be added to monitor the amperage use. This is handy when transmitting, as it shows a noticeable rise when the PTT switch is depressed. You can forgo meters altogether, and make the unit just a simple parallel wiring arrangement. All that is a matter of choice.

This device is one to assuredly keep in mind. If all else fails, find someone like Mike to build you one. You won't be sorry you did.

Computer-Controlled Radios

Let's turn our attention to computer-controlled radios. This is a relatively new area that is catching on quickly. These rigs are basically all-mode (AM, FM, CW, SSB, etc.) scanners that connect to your computer through a serial or parallel port. The guts, or electronics, are enclosed in a small box that usually has a telescoping antenna, but the computer, and its monitor, does most of the work.

For example, in addition to telling you what frequency it is receiving, it can also show S-meter readings, frequency shift and other pertinent data. Coverage is often from around 10 megahertz to as high as 1.3 gigahertz and can be scanned in bands, specific frequency lists or the full spectrum. Unlimited memory capacity is available due to the large memory of the computer, and most companies offer on-line help service.

Hence, these gems can be an excellent choice for a DX or scanner type receiver especially if you spend a lot of time at a computer. They can run in the background to be brought up at will. If and when you decide to add a full-feature receiver, keep this option in mind.

Portable SSTV

This next one is one of my favorites. I have a fascination with sending images over the air, so this is right up my alley. What I'm referring to is portable slow-scan television (SSTV). As of this writing, one company (I won't mention names, but they use "TS" a lot in their radio designations) offers a completely self-contained SSTV accessory for use with its high-end HT.

This thing is slick! In one package you get a charge-coupled device (CCD) color TV camera, a liquid crystal display (LCD) monitor and an image scan converter. With the unit attached to the HT, it records a still image, converts it to analog audio that can be transmitted and then converts any return signal to a digital format and displays it on the LCD.

What more could you ask? Now, now! Let's be reasonable! If you are like me (interested in SSTV and ATV) this device is a dream come true. I mean, I have got to have one. And, if all of you out there buy enough of these books, I will be able to afford one!

GPS

Here is another interesting area, although I have to confess I haven't done much with it, just a few demonstrations. It is called the Global Positioning System (GPS) and is a really nifty way to tell where you are. It also tells you where to go. NO, I don't mean like that; where to go if you are traveling somewhere.

As I understand it, this little satellite receiver knows where you are, and if you want to get to point A from there, you tell it where point A is, and it tells you the best route to take to get to point A. HUH? Yeah I know, that did get kinda breezy. Let's try this.

The GPS receiver picks up its bearing from the satellite that determines the receiver's location. From this information, the GPS can determine where other locations are in reference to its own position. Using that data, some of the units can lay out paths to the second location.

These units also give you other information such as the elevation at your position, so they can be very handy to have around. I'm also told they are extremely accurate which makes them a natural partner for APRS stations. In this situation, the GPS is interfaced with the computer and/or radio to add further information regarding the APRS station.

GPS receivers come in a variety of different complexities. The most basic unit, thus cheapest, displays a compass and other data on an LCD screen, while the most elaborate receivers can actually display road maps, even in color. Naturally, price increases accompany added features.

Well, that's about all I know, or could find out, about GPS. It does sound fascinating, though. Several companies are making GPS receivers, and I do know the US military has been making good use of GPS ever since Operation Desert Storm and loves the system. If it's good enough for Uncle Sam, well....never mind.

Books

On a last note concerning miscellaneous items, don't forget the massive volumes that have been written about amateur radio. I won't recommend anything specific (not even my own work), but many really good books have been produced to help you with this hobby. Take a little time to explore this aspect.

For every subject I have covered there is a book written concerning it. Many questions you might have can be answered by reading these texts. As for availability, all the major HAM distributors carry excellent lines of HAM-related books. Also, many of the electronics/electronic part companies also handle some of this information.

Conclusion

Wow! I don't know about you, but I'm exhausted. It's a nice exhaustion, however, as discussing amateur radio equipment is always a pleasure for me. There is so much to look at, consider, and talk about out there that it can be a little overwhelming at times. I know I felt that way at first.

I soon began to get a handle on most of it, and then the fun began. One really nice product of all those companies competing for your money is that prices seem to be coming down all the time. HAM radio absolutely can be an inexpensive hobby. A lot more so than some others. I know a guy who spent nearly $3,000 on a set of golf clubs. Now, I don't have anything against golf, but he still couldn't break 100 on the course. He sure could have equipped a dandy HAM shack with same $3,000.

Anyway, I hope you have enjoyed our journey through the world of HAM radio gear. I enjoyed bringing it to you. Next, we look at the many activities this hobby provides.

HAM Radio Activities

Introduction

Now that you have your license and equipment, what is next? Well hold on to your seat ace, I'm going to tell you. All of the things you can do within the HAM radio realm are amazing. We talked about the areas of the hobby earlier in this section, so let's put that discussion into high gear.

As I'm certain you noticed in that first segment, there is a lot more to amateur radio then just sending voice or code over the air. That is a big part of it, but depending on your interests, HAM radio offers many avenues from which you can expand the basic hobby.

QSOs

Pronounced "Q-So", the QSO is perhaps the essence of amateur radio. In short, a QSO is a contact, or conversation, between two HAMs. It can be by phone (voice) or code (CW) and represents the purest form of HAM radio. It is where we all start. QSOs are also downright FUN!

The longest range QSOs are made on the HF bands, but we have covered this before. To review, the characteristics of those frequencies are such that they lend themselves well to literal around the world communication. Hence, many of us want to learn Morse code so we can upgrade our licenses and get on the HF bands.

Until you do, the VHF and UHF spectrums have a lot to offer. In most areas, QSOs are numerous on the 2-meter and 70-centimeter bands, and in some locales, activity on 1.25-meters, 33-centimeters and 23-centimeters is quite lively. All of this depends on the size and enthusiasm of the local HAM community.

With 2-meter, for instance, communication is normally conducted in one of two formats: simplex or duplex. Simplex is an unassisted contact between two stations on a single frequency. In my area, many of the local HAMs gather on 147.555 megahertz, also known as the triple nickel. Here using FM, the conversation is lively during the day and into the evening hours. (Usually, everybody goes to bed around 11 p.m.)

Every Friday at 10 p.m., the world-renowned "Hootie Owl" net is held on the triple nickel. This is an informal network that features the "net manager's" opinion on some subject, and the famous "atomic drop" award for the HAM who has demonstrated, during the past week, a need to be "lifted to a height of 147.555 feet and dropped on his head." Actually, this can be a good or bad award. The Hootie Owl net is a bit unorthodox, but great fun for all who participate in it.

As can be seen, this frequency provides a meeting place for HAMs wanting to get involved in one of the canons of amateur radio, yackin'. I have been told by HAMs who have worked other areas of the country that these simplex talk frequencies are quite common throughout the United States. It seems that 146.520 megahertz, which is 2-meter's national FM calling frequency, is popular. So, if you are traveling, keep an ear on that spot.

The other form of local communications involves duplex, which is the use of a repeater to make the QSO. Since I have discussed repeaters at length, I won't get back into that. This method is highly accepted among HAMs on the VHF and UHF bands.

In short, the QSO is the foundation on which all of HAM radio stands. From its beginning, this hobby has evolved from getting on the radio and talking to other amateur radio operators. A fine tradition indeed!

DXing

For many HAMs, this is where it's at. Searching the airwaves for distant and new stations to communicate with. This activity has a dedicated following, especially among HFers.

It does involve keeping logs of the stations contacted, which is part of the fun. Many HAMs have literally stacks of logbooks, some dating back many years. It also involves sending and receiving QSL cards. What is a QSL card? It is a written confirmation that the contact was made. Usually, it has all the pertinent data and some get very fancy, with pictures and the like. Again, many older, more experienced HAMs have stacks of these.

Additionally, DXing can include awards (wall paper) issued by various HAM-related groups (the ARRL for one) for such things as contacting all states, or contacting all states on 20-meters, etc. For serious DXers, this becomes an inherent part of the excitement. Many attempt to earn as many awards as possible, and it is evidence of their DXing expertise. The awards can literally wall paper their shacks.

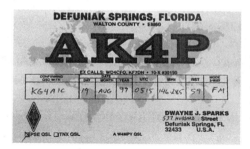

A typical QSL card. This one is from Dwayne AK4P from Defuniak Springs, Florida.

Classic QSL cards featuring some of the many designs from past years. Courtesy of Don W3MK.

Classic logbooks from the 1940s. These have covers of different colors to assist the user with organization. Courtesy of Don W3MK.

Two typical types of logbooks: a spiral-bound log and an ARRL log sheet for a three-ring binder.

Propagation

While this is not an activity as such, I felt it might be a good idea to briefly discuss this topic here. That is because propagation is the key to all of what we have just been talking about. Well, at least it plays a very important part.

By definition, propagation encompasses all the atmospheric conditions that either aid or deter the radiation of radio waves. These include such things as weather, sun spots, atmospheric ionization and a host of other stuff.

The ionosphere is divided into layers, such as D, E, F1 and F2, that have different effects on different frequencies, and these layers appear and disappear according to the time of day. For example, the D, or lower layer, needs the energy of the sun to exist, thus it vanishes at night. The F layers, however, are present at all times and have a major influence on sky wave communications. They are the highest layers, and combine into just a single F layer at night.

Heavy ionization caused by sun spots is another factor contributing to good propagation (signal carry). When the upper layers ionize, they do not allow as much radio energy to escape into outer space. Hence, that energy is reflected back to earth. Naturally, that improves not only the distance signals travel, but also the strength at which they arrive.

As I said, this is a brief dialogue concerning propagation, as this subject could easily be a book in itself. Howeverm you will find a more detailed discussion in Section 3. Once you get established in HAM radio, I suggest you take a more detailed look at propagation. Such exploration helps you better understand the conditions that affect the range of your radio transmissions.

HAM Radio Clubs

Okay! Enough of that mind provoking sh...uh, stuff! Let's get into something more interesting like HAM radio clubs. If you have two or more HAMs living in one location, the betting is your also going to have a HAM radio club. It's almost a tradition! It's a good tradition too.

Clubs dedicated to amateur radio are not only great fun for the HAMs, they can also be of monumental value to the local community. Amateur radio groups donate their time and equipment for local walk-a-thons, bicycle races and other events where short-range communication is both helpful and a safety factor. By positioning HTs and/or mobiles along the events route, progress can be monitored and medical help can be summoned if needed. Once again, amateur radio becomes a service to the community. I find very few HAMs who don't take pride in that.

Another nice aspect of being a club member are the group get-togethers such as breakfasts, luncheons and dinners. These events provide an opportunity for local HAMs to conduct eyeball (face-to-face) QSOs. They can be very enjoyable occurrences with a lot of amateur radio talk. Besides, next to talking on their radios, most HAMs love to talk in person, especially about the hobby.

Most clubs also produce a newsletter that is very informative. Most of them try to be. If you have an interest in this area or background in publishing you are no doubt of substantial assistance to whomever does the newsletter. This is a way for you to contribute to your local club and can be really fun.

Perhaps you should talk with the club's newsletter editor. I have a feeling he or she will greet you with open arms even if you don't have any prior experience. You can always learn! Since HAM radio is a hobby, many people don't have a lot of extra time to bestow upon the club. Usually, any offering is greatly appreciated.

There are many other areas clubs are frequently involved in. I discuss some of them as we go along. So, if you have a special interest, it is probable that someone else shares your interest. Also, you are likely to find an established club activity that catches your attention. In short, I highly recommend joining one or more of the clubs in your area.

HAMfests

With most amateur radio clubs, this is the big event of the season. These take a full year to plan, but boy are they worth it! At HAMfests, you have the opportunity to see new and used gear, meet new HAMs as well as old friends and generally immerse yourself in a full day, or even two, of HAM RADIO! There is nothing like it!

I try to attend every HAMfest in my area, as they are an experience all their own. They provide a chance to obtain equipment you need or want for your station. This is nice, as it's all well and good to look at a picture and description of what you plan to buy in a catalog, but there is no substitute for being able to get your hands on that item. You know, hold it, look at it, just get a feel for it. That way, you know exactly what you are buying. Try it, I am sure you'll like it!

An equipment-rich display at the 1999 Huntsville, Alabama, HAMfest.

Hamfest flyers that are sent out to prospective attendees. Flyers for other fests are available at most shows to advertise upcoming events.

Need an oscilloscope? There were a few available at the 1999 Huntsville, Alabama, HAMfest.

In addition to this facet of the HAMfests, there are the seminars, workshops and testing opportunities that accompany many of the events. Here, you have a chance to sit in on, and participate in, discussions involving HAM radio. These might include reports by the state repeater conference, ARRL meetings, Skywarn, ARES and other emergency organization symposiums and of course, an opening to take VEC tests for licenses and/or upgrading.

All of this adds to the utility of a HAMfest. Naturally, there is always something to eat. Some clubs have barbecues, others have catered food, and some even hold dinners. You won't starve if the HAMfests I have attended are any example.

Check the listings in QST Magazine (the ARRL monthly publication devoted to HAM radio) or your local club(s) for upcoming HAMfests in your area. Even if you have to drive a distance to attend, I believe you will find the experience well worth the trouble.

Swap Meets

Akin to the HAMfest is the swap meet. This is much like an amateur radio flea market where mostly used equipment is the offering. However, don't sell these short. A lot of really good bargains can be found at swap meets.

One caution though, and I don't want to dwell on this, but it bears mentioning. In the way of an old axiom, "let the buyer beware!" Be sure you carefully check out anything you purchase at such an event. The vast majority of the dealers are honest, reputable folks, but like pirates, some crooks do exist. Once they are gone, good luck finding them again.

As I said, check the stuff out before you buy it. If a dealer refuses to demonstrate the functionality of the gear, chances are it doesn't work. So look elsewhere.

Otherwise, swap meets are a real day at the beach. They are not as prevalent as they once were, but when you find one, be sure and stop in just to see what is on sale. I always do, but then, I'm a sucker for that sort of thing.

Foxhunts

Here is a club event that is easy to organize and a real treat to participate in. The object here is to have one member of the club (the FOX) hide somewhere with a radio, and the rest of you go out and find that member. This activity is not only entertaining and challenging, it is also educational.

You learn a whole bunch about radio behavior from this affair. I kid you not! Such things as signal bounce and hot spots take on a new meaning when you are out there trying to pin the location of the hidden transmitter. Perhaps as important is the training you will get in direction finding (even though you won't realize it is training at the time).

This aspect can come in very handy when someone has a stuck microphone and needs to be located to inform the individual of the problem. That situation can tie up a repeater for hours without the culprit even knowing he or she is the one.

Also, I'm sad to say that from time to time, illegal operators (pirates) and/or offensive operators are encountered that need to be located. Fortunately, this doesn't happen much in most HAM communities, but it does happen. That is when the foxhunt training, you know... that stuff you didn't know you were really learning, comes in to play.

(L to R) Randy KV4AC, Fred K8AJX, and Rik KU4PY use the directional antenna to look for the hidden signal at a foxhunt.

(L to R) Rik KU4PY, Fred K8AJX and Randy KV4AC take a "fix" and plot it on a map in order to find the hidden transmitter (the fox).

The skills learned from an adventuresome event come back to provide the needed talents for the job. Several competent foxhunters can usually track down the offending station in short order. Then, we take him out, give him a fair trial and hang him. No!!! Not really! I'm just kidding. We simply turn him over to the FCC! That often eliminates the problem.

Field Day

Each year, Field Day, which is always held the last weekend in June, is the largest contesting event sponsored by the American Radio Relay League (ARRL). Virtually all amateur radio clubs have some degree of involvement in this competition, as well as many individual stations.

A vehicle used for CW and Satellite stations on loan from the Alabama EMA. Note the generator at the rear.

Field Day is a 24 hour function where the radio equipment is usually taken to a remote location and operated off an alternative power source. This could be solar panels, a wind mill, a water movement generator but is normally a gasoline powered electric generator. I say usually, as certain categories of the contest allow for standard 120 VAC household power.

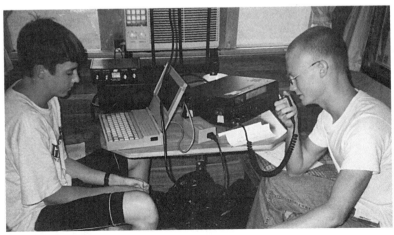

Drew KE4MDS watches as Mike KG4CSL works the 10 meters at the Novice station.

The object of this game is to make as many contacts during the 24 hour period as possible. In that respect, it is much like most other contests. These contacts are logged, submitted to the ARRL and your station is ranked accordingly.

Each of the setups is given a specific call sign, often the club's call sign, and treated as one station. They are also placed into classification dictated by the number of transmitters operating. For example, if you have three different radios transmitting and you are a club, the station classification would be 3A.

Some categories, like a Novice station, do not count as a separate transmitter and thus do not affect the overall station designation. Also, extra contest points can be earned by such things as operating a solar power station, having a greeting center, making satellite contacts and the like. All of this gets pretty complicated, so if you would like more detailed info, contact the ARRL either by phone, snail mail or their website - www.arrl.org.

Field Day is great, though. This year, I was railroa....uh, given the opportunity to serve as my club's Field Day Chairman, and had a good time with the project. It required a lot of preplanning and legwork, but the event really fell together, making it enjoyable for everybody involved. Try not to miss this when it comes around.

Networks

Networks, or Nets are a radio gathering of local, and sometimes not so local, HAMs with a special interest. There are 2-meter nets, satellite and weak signal nets, emergency organization nets such as Skywarn and ARES, conversational nets and nets that cover just about every individual area of amateur radio.

While many nets are held on 2-meters, you also find them in the HF region. 10- and 80-meters are favorite bands, but networks are heard in other areas as well. If you do have a special interest, checking into a network devoted to that concern is a great way to receive information and meet other HAMs with the same interest. Usually, checking in amounts to nothing more than keying the microphone and giving your call sign, name and maybe the county you are from. Make sure you wait for the appropriate time (when asked to check in) and that you listen for other stations trying to do the same. With busy nets, it is easy to walk on another signal.

In my area, I'm involved with the National Weather Service (NWS) Skywarn organization. Each week we hold a net concerning information about Skywarn and severe weather spotting. On a normal night, around 25 stations check in, and we discuss a variety of subjects ranging from Skywarn related events to general weather data and training.

These nets are conducted (called) by a net controller whose job is to maintain some semblance of order no matter how rowdy the crowd becomes. No, I'm just kidding again. At least with Skywarn, everybody behaves quite well, and the nets are very productive.

For my local Skywarn, I'm one of four people that calls the net on a rotational basis. We fall under the net manager, who is the local chapter director. His duties also include acting as one of the four weekly net controllers. That is just the way it is handled in my area. While an unofficial arm of the National Weather Service, Skywarn is left pretty much to the local participant in terms of how it is managed. As long as we do our job, the NWS is happy.

That job involves bringing up the network during severe weather conditions and providing information to both the local area and the National Weather Service. Whenever the NWS issues a weather warning for our six county area of responsibility, Skywarn comes up. We then receive information from the many weather spotters in the field, and pass this information on to people listening to our transmissions.

You don't have to be a HAM to receive this info. Anyone with a scanner that monitors HAM frequencies (as far as I know, that is just about all of them) is able to hear us. For example, most of the local TV stations listen to the Skywarn net during severe weather. We normally use one of the club repeaters for our communications, so the signals are clear and strong.

So, once again, HAM radio answers the call as a service to the community. Skywarn is just one of many emergency management groups that use amateur radio in this fashion. In times of need, HAM radio has been there to aid anyone who has needed help. With regard to communications, often during major disasters amateur radio is the only game in town.

Hence, here is an activity you may want to look into. It may give you a chance to be of assistance when your assistance is most required. The NWS sponsors regular Skywarn training classes that again are free and very educational. These classes train you as a severe weather spotter, which not only benefits Skywarn, but could also save your life or the lives of your loved ones and others.

Naturally, I'm partial to the Skywarn program. I have learned a lot from it and feel a real sense of accomplishment in my commitment to Skywarn. Thus, I encourage any new HAM, or old-timer, to consider getting involved. My area is lucky to have great participation, especially during times of severe weather. This has been highly advantageous to local residents, as well as area officials.

Educational Efforts

Another way to help your community through HAM radio is to get involved with school programs. Many clubs have volunteers that visit junior high and high schools, in some cases even elementary schools, to encourage student to become HAMs. This effort not only benefits amateur radio, but also benefits the students.

It provides a hobby and educational experience for those with an interest in communications and/or electronics. Often that hobby blossoms into a profitable career. Even if it stays just a hobby, it offers younger members of our society a pastime that is productive and rewarding. It is something they can enjoy for the rest of their lives.

As for amateur radio, these programs bring us fresh talent in a hobby that is always looking for new blood. All of us in HAM radio have someone we recognize as the person responsible for our being here. This person is affectionately known as our Elmer. For many HAMs, that title has been bestowed upon them by entire classes of students. That must be a warm feeling, knowing you have been able to help that many kids find the hobby and perhaps stay out of trouble along the way.

It isn't a difficult task, and most schools are overjoyed to have you come and illustrate HAM radio. A little equipment, a gift of gab and you're all set. Many clubs gladly appropriate the funds necessary to purchase and put together a demonstration station, as this is a very positive service to the community. Don't forget the FCC's third-party rule either. As long as you are present as the station's control operator, the kids can talk their heads off on the air. That usually provides a real incentive!

So, if your club doesn't have such a program, help it get one started. If on the other hand it does, get involved. This may be the ticket for you. At the very least, it is a rewarding way to spend some of your spare time.

Contesting

Contesting, what do I say about contesting? It is big with a lot of HAMs and involves hundreds of hours of their airtime. Organizations like the ARRL sponsor a veritable smorgasbord of different contests throughout the year that furnish all kinds of fun, excitement and awards for those who participate.

That is just the beginning. When I last visited the ARRL contest section on their Web site, I counted a total of 168 contests listed from October of 1998 to September 1999. These are not all ARRL contests by any means but, they are out there!

However, I have found there is little gray area regarding contests. HAMs either love them, or well I won't say hate them, but they don't like them too well. At least that seems to be the consensus I have gathered thus far.

Contesting is something you will have to try for yourself and see if you like. I have been involved in a number of contests and found them to be a great deal of fun. If you are into DXing, then contesting has to be a natural for you.

As with so much of this hobby, contests come in a variety of colors, flavors, sizes and shapes. There does seem to be one common denominator and that is to make as many contacts as possible. I've yet to hear about a contest that gave awards for the fewest contacts. This ain't golf!

Some of these events involve specific modes such as SSB or CW, others seem to want contacts from certain areas and then there are the ones that want all contacts on a single band. This goes on and on, and I suspect there is a contest out there for just about everybody. There has to be.

Anyway, here is an area that definitely deserves consideration. You may not like it, but then you may find that this special arena of HAM radio is just right for you.

Balloon Launches

Now, this is for real. You are going to hurt some feelings if you make fun of this one. Take my word for it! Launching radio-equipped balloons is a major project with many amateur radio clubs. If you don't believe me, check out the cover of the January 1999 QST magazine.

Think about this for a moment. Your club gets together and sends up a helium-filled balloon. The payload for that balloon consists of HAM radio transceivers for various bands (VHF and UHF are recommended), an ATV system and some weather instruments connected to a radio downlink. With that stuff on board, you are able to see where you're going, your balloon that is, talk to HAMs that are within range of the balloon and tell what the weather conditions are at whatever altitude your balloon attains.

Not bad, when you think about it. If fact, this is a pretty impressive club project. As for altitude, it is not at all unusual for these balloons to reach 100,000 feet. That should give you one heck of view of the ground! And, think about the range on the VHF/UHF radios at the height.

Also, if you want to get real ambitious, install an APRS unit in the payload. This addition allows the APRS network to track your balloon. It comes in handy when it's time to recover the equipment.

Now this is not the cheapest project the club can undertake, but by same token, it isn't that expensive either. Surplus weather reconnaissance balloons can range in price from $10 to $100 depending on how well you shop. A protective case for the equipment/instruments must be constructed as the temperature can reach 60 degrees below zero up there!

With a little ingenuity (something HAMs are famous for) and teamwork, your club could have one of these off the ground and running. It might not be a project that is undertaken every year, but it sure could be fascinating and educational, even if it was only done once.

So, next time the club starts pondering a new venture, you might bring up a balloon launch. After the other members stop throwing stuff at you and give you a chance to explain, they might just see it your way. You never can tell!

Recruitment

It is the sworn duty, a sacred oath, for all HAMs to recruit new blood into the hobby. There is no room for argument here! No gray area, no excuses, no way out! Once you become a HAM, you must take on this responsibility without question or complaint!

Well, it isn't quite that bad! Perhaps the above declaration was a little strong. For the sake of the hobby, it is a good idea to encourage even a slightly interested party to pursue amateur radio. Now that doesn't mean you have to break out the torture rack and thumb screws. Be nice, but be firm!

Drew KE4MDS and his "Elmer", otherwise known as Drew's dad, Fred KR4YK.

Explain the attributes of the hobby, the camaraderie, the fellowship (Oops, I kind of repeated myself), well, you get the idea! Talk up HAM radio, let that person know how exciting and rewarding this venture can be and above all, be nice!

Okay, okay! Enough of the razzle-dazzle! We don't want to scare anybody off either. In all seriousness, amateur radio is one heck of a hobby. For those of us involved, it is hard to understand how anyone could not find this field enchanting. I for one have grown to love HAM radio, and I find that same feeling among many fellow HAMs (although, sometimes they don't want to admit it).

I have also found that once people begin to understand all that amateur radio is, it takes on a whole new meaning for

Tom WB2ZKA became interested in HAM radio through his father. Tom encouraged his wife to try it. Now Diane N2UEC is a licensed Technician Plus.

them. Often, they do become interested. This newfound interest frequently surprises them, as they never saw themselves pursuing anything that involved radio or electronics. Don't lose the faith. There are more prospective HAMs out there than you might think. Who knows, it may be you that brings someone new into the hobby. Then we will all be able to call you Elmer too.

HAM Traditions, Etiquette, Conventions, Safety, Do's and Don'ts, etc., etc., etc.

Introduction

Nah, don't worry. I'm not going to lecture you! This is simply a brief section on some of what makes for a good HAM. Good in the sense of proper operating skills and courtesy to others. Additionally, it fills you in on a few of the hobby's time-honored traditions, and all of this will make your experience with amateur radio just that much more enjoyable.

I also cover safety and what you should do and not do when it comes to that subject. In reality, HAM radio is a very safe hobby when you compare it to sky diving, bungee jumping or golf. And, if some good old common sense is used, you should never have a safety problem. For the sake of the old adage, "better safe than sorry," I will go over this.

Regarding courtesy, I'm not saying you aren't a courteous person. Not at all! I'm sure everyone reading this book is courteous, but with HAM radio, sometimes you're being discourteous and don't even know it. When I first got on the air, I was razzed about using the 10-Codes (you know, 10-4, 10-8, etc.), but I didn't know they were considered taboo on amateur radio.

The courtesy part is just to save you a similar fate and let you know what is considered courteous and what is not.

With that said, let's dive into this. I keep this brief, since I feel most everyone catches on quickly. If you have any questions, don't hesitate. Just raise your hand!

First Names Only

One of the truly nice aspects of amateur radio is that everybody uses his or her first name. On the air, I'm Carl KG4AIC, not Mr. Bergquist! We do have a guy in our area, KR4WN, that we call Mr. Rice, but that is because his first name is Jim and we alreay have several Jims. Hence, for clarity sake, it is easier to call him Mr. Rice.

Except for unusual circumstances like that, HAMs generally refer to each other by their first names. Don W3MK was telling me about contacts he made with King Hussein I of Jordan. As you may know, King Hussein was an ardent supporter of HAM radio, and on the air he was Hussein, not King Hussein.

United States Senator Barry Goldwater was another devoted celebrity amateur radio operator, and on the air he was Barry. This is an established tradition with deep roots among HAMs. It is one that traverses virtually all boundaries and protocols. You will enjoy it, as not only does it strengthen the fellowship among HAMs, but it also makes operating much easier. Personally, I have a hard time with last names!

Camaraderie

Webster's New World Dictionary defines this term as, "loyalty and a warm, friendly feeling among comrades", and I think that pretty well describes the atmosphere within most of the amateur radio world. Personally, I find a benevolence and generosity that is hard to match. Perhaps this helps explain the devotion so many HAMs have for their hobby.

With any subculture in our society, there is bound to be a dedication to the precepts of that group. However, with amateur radio the allegiance seems particularly intense. The answer to this lies in the conviction HAMs share toward the prosperity and preservation of something they truly love.

At the risk of sounding like an amateur psychologist, that type of commitment normally builds an extremely robust relationship between two entities. I have seen evidence of this in the HAM radio community. HAMs and their hobby are, for the most part, inseparable.

There is a social aspect of HAM radio that has been its heart and soul since day one. This led to the unwritten rule HAMs follow in their interaction with each other. For new HAMs, the rule is often not obvious. To clarify this, I asked Stephen Mendelsohn W2ML, First Vice President of the American Radio Relay League (ARRL), his opinion. His answer, "every HAM helps another HAM. That is the unwritten rule!"

Definitely words that HAMs SHOULD, and most DO live by. In fact, for most this is an essential role of the hobby! I think you too will recognize this aspect of amateur radio. In the end, it becomes more than just a hobby. It becomes an important part of your life.

General Courtesies of HAM Radio

For this segment, let me cover some of the areas that fall into this courtesy thing. Of course, this is a partial list, but I think it encompasses the most important rules that HAMs live by. The finer points are not emphasized as much, and you will pick up on those as you go along. And, usually you can do this without stepping on too many toes.

Appropriate Language

I'm almost ashamed to include this one, but if you have monitored the Citizen's Band (CB) lately, at least certain channels, there are individuals that don't seem to understand this point. Please use language that you wouldn't mind small children hearing. I suppose that applies to larger children as well.

Profanity and lewd and suggestive conversation is not deemed appropriate for HAM radio. In many cases, it is also illegal. The FCC not only takes a dim view of it, they have the power to make your life miserable if they feel you are violating their rules and regulations.

There just might be young ears listening! Mark KQ4SX shares his hobby with his family, here with son Steven, at his living room station.

I'm not going to dwell on this, as I think, for the most part, it isn't necessary. Just be aware of what you're saying when the PTT switch is depressed. I know, an occasional word may slip out, but be careful what that word is. Some are viewed forgivable, while others are not. Big Brother has spoken!

Operating Procedures

Most operating procedures involve the same old common courtesy we have already discussed, but here are some tips that may help you get acquainted with, and used to, on-air courtesy. When dealing with radio, it is easy to forget yourself and blunder, especially when you're first starting out. If you keep these methods in mind, you may blunder less often.

First, listen before you transmit. I know that in a haste to get on the air, it is easy to just pick up the microphone and start talking or begin pounding brass. However, if you don't listen first to make sure the frequency is clear, you are likely to step on someone. This is not good! HAMs are quick to reprimand, or worse, ostracize other HAMs that frequently violate this rule. Additionally, this can be considered QRM, or malicious interference, which could get you in hot water with the FCC.

Simply take a moment to check the frequency before you send a signal. There are certain wavelengths where it is so busy you may have trouble sending, but they are few and far between.

Along these same lines, allow a brief one- to two-second delay during QSOs before replying. This allows other stations to join the QSO, or ask for a break in traffic. Again, in the excitement of a conversation, it is easy to want to immediately reply to the other party but, HAM courtesy dictates that you provide those short breaks. It makes you a better operator.

Also, remember that no one owns a frequency! The spectrum is allocated to the amateur radio service and the HAMs that operate there. Try to be considerate to others, and they will do the same for you.

Regarding frequencies, this no-ownership rule applies to other situations as well. I mentioned the networks that many clubs hold on the air. These are often conducted over repeaters for the obvious reason. If it comes time to start the net, and someone is using the repeater, postpone the start until that person is finished.

If the individual is persistent and doesn't want to clear the frequency, a polite mention that a network is about to start might be in order. A vast majority of the time, this does the trick. He/she may not be aware of the network and/or its starting time.

If use by the other party still persists, then he/she is violating the courtesy rules. However, it is far better to move to an alternate frequency than try to argue that person off the air. The bottom line here is that individuals have just as much right to that frequency as the net does.

Concerning our next subject, it is one that plagues many voice QSOs. It is also something we all do from time to time, and that is we do not speak clearly. So, make an effort to enunciate your words when you are on the air. Also, and this is one I have to watch, don't speak too quickly. I know the FCC likes brisk exchanges, but don't overdo it. Slow down your speech. It makes you much more popular with other HAMs.

Along these same lines, it is often hard to understand distant stations, or even close ones for that matter. This can be the result of intermod, other interference or just plain sloppy speech. So, try to use as many phonetics as possible, especially when giving names and call sign. I have included amateur radios' accepted phonetic alphabet for your convenience, so look it over and use it when necessary. It isn't hard to learn. I wish code was as easy! Again, no 10-Codes! Remember, this isn't CB!!!

FCC and Amateur Radio Band Plans

With this area, we are dealing in two different, but related topics. The first is the FCC divisions of the various bands according to license class, and the second involves HAM tradition. The relativity lies in the fact that both establish canons for how each band should be partitioned.

Regarding the FCC, they have set up mandates as to what portion of a band each license class can use, and for what (CW, Phone, etc.). These, of course, are law. For example, a Tech Plus is allowed to use the 28.10 to 28.50 megahertz portion of the 10-meter band, while a General class licensee can use all of 10-meters (28 to 29.70 megahertz). In a similar example, Advanced HAMs have CW privileges of 3.525 to 3.750 megahertz in the 80-meter band, while Extras can use the full spectrum of 3.500 to 4.000 megahertz. This is done primarily to encourage license upgrading but does reflect the FCC's version of band plans.

The amateur radio version of a band plan involves the segregation of each band into specific frequencies designated for specific operation. For example, in the 2-meter band, 144.10 to 144.20 megahertz is assigned to EME and weak-signal SSB, while 144.60 to 144.90 megahertz are set aside as FM repeater inputs.

146.40 to 146.58 megahertz are simplex frequencies, and 147.00 to 147.39 are repeater outputs. Similarly, on 70-centimeters, 438 to 444 megahertz are reserved for ATV, while 435 to 438 are satellite only. In this fashion, each band is better managed, thus more efficiently used.

However, unlike the FCC edicts, this second band plan is strictly voluntary. There is nothing legally binding about these frequency designations, merely a hope and desire by the amateur radio community, in general, that they are followed.

So, you may elect not to follow them if you wish, but I have a feeling you won't be too popular out there if you do. These band plans have served HAM radio well for many years and are, for the most part, taken as law. You will get some static if you decide to use frequencies for purposes other than specified. This I can promise you!

Anyway, I'm confident you wouldn't do anything like that. Wait a minute, there's a question in the back. No, I guess not. Must have changed his mind. All joking aside, the amateur band plans do work well and do keep a band like 2-meters organized. I think you will like them.

Safety for RF and General

You remember RF Safety, don't you? He used to live down at the end of the street. Had a brother who went in the Army, and now he's General Safety. Uh Huh! You see, I did that because I used to have this public speaking professor who said you should always start off with humor. Obviously, he was wrong!

Anyway, perhaps it was inappropriate of me to begin this subject with a joke, as this is not a laughing matter. I don't want to frighten anybody as that is not the purpose of this section, but safety does bear mentioning. For the most part, HAM radio is quite safe, but as with most anything, certain precautions should be observed. It is those precautions that we touch on here.

First, let's talk about RF safety. I heard he got married. Ha, Ha, Ha. Nah....Never mind. I'll straighten up in a minute here, just bear with me. Okay! RF safety. Radiated radio frequency energy can be hazardous to your health if certain conditions are not observed. Much like microwaves can boil water and heat food, all RF energy can cause the heating of body tissue. Microwaves are just more efficient at it.

Due to this danger, federal guidelines and restrictions have been established concerning RF radiation. This involves the Maximum Permissible Exposure (MPE) allowed for any given transmitter/antenna arrangement. The guidelines are in the form of tables that indicate what the MPE is for that station.

Included here is the frequency, type of antenna and power output. The last item is calculated by taking a radiation measurement, usually with a field-strength meter (FSM), and applying that information to the formula: mW/cM2. This equates to milliwatts per square centimeter, and the station must fall within the maximum permissible level to be deemed safe.

I realize this is a little complicated, and if your station puts out less than 50-watts, you won't have to be concerned with it. The FCC does not require RF monitoring and routine evaluations on transmitters under 50-watts. If you are transmitting 50-watts or more, as most HF rigs do, then you are required to survey your station annually to determine if it is RF safe. This is called a routine station evaluation.

Actually, this is not as difficult as it sounds. With the reference tables previously mentioned, it is merely a matter of taking an FSM reading and checking it against the table. If your reading is at or below the MPE, you're in good shape.

Table MPE - 1

Limits for Maximum Permissible Exposure (MPE)

Frequency Range (MHz)	Electric Field Strength (V/m)	Magnetic Field Strength (A/m)	Power Density Strength (mW/cm\2\)	Averaging Times (Minutes)
(A) Limits for Occupational/Controlled Exposures				
0.3 - 3.0	614	1.63	*(100)	6
3.0 - 30	1842/f	4.89/f	*(900/f\2\)	6
30 - 300	61.4	0.163	1.0	6
300 - 1500	...	...	f/300	6
1500 - 100,000	...	...	5	6
(B) Limits for General Population/ Uncontrolled Exposure				
0.3 - 1.34	614	1.63	*(100)	30
1.34 - 30	824/f	2.19/f	*(180/f\2\2)	30
30 - 300	27.5	0.073	0.2	30
300 - 1500	...	...	f/1500	30
1500 - 100,000	...	...	1.0	30

f = frequency in MHz

* = Plane-wave equivalent power density

Note 1 to Table MPE - 1:

Occupational/controlled limits apply in situations in which persons are exposed as a consequence of their employment provided those persons are fully aware of the potential for exposure and can exercise control over their exposure. Limits for occupational/controlled exposure also apply in situations when an individual is transient through a location where occupational/controlled limits apply provided he or she is made aware of the potential for exposure.

Note 2 to Table MPE - 1:

General population/uncontrolled exposures apply in situations in which the general public may be exposed, or in which persons that are exposed as a consequence of their employment may not be fully aware of the potential for exposure or cannot exercise control over their exposure.

The height of the antenna highly influences the FSM reading you get, so for the most part, many HAMs are not going to get measurements that are illegal. We keep our antennas up so high, they are at distances greater than specified by the MPE tables. If your power output falls within the regulated level, be sure to check and record the result once a year. If the FCC comes calling, they are going to want to see that data.

On a last note about RF safety, keep the hazard of radiated RF uppermost in mind at all times. It can shock and/or burn you and others badly. You don't want to tarnish your affection for amateur radio with a nasty incident involving RF energy.

Antennas

One of the true mysteries of HAM radio is antennas. They are a science within themselves, and can be a fascinating part of the hobby. But, they....don't you just love it? There is always a "but". Anyway, they can also be dangerous, if not handled properly.

Aside from the RF danger, just the process of putting them up or taking them down can place you in jeopardy. It is a good idea not to do either alone. If they get away from you while being raised or lowered, the mast/antenna assembly can fall on you or other objects (like your roof). That can lead to serious injury, or the need for a new roof.

Known as "tower rats", many HAMs enjoy climbing towers for installation and repairs.

When it comes to towers, they are great for their easy access to the antenna(s). You can climb them and be right where you need to be to install, remove or repair the antenna. You can also fall off the tower. I won't get into what kind of damage that might do to you.

Anytime you climb a tower, be sure you wear and use a safety belt. These are heavy leather waistbands that can be secured to the tower. Safety belts allow you to attach yourself to the structure and have your hands free to work. In the event you slip, the belt catches you, and you won't end up on the ground, a crumbled, broken mess.

Incidentally, for those left on the ground, it is an excellent idea to wear a hard hat. This helps prevent major damage if some clumsy fellow HAM drops a tool, or possibly the antenna, on your head. Now, I know some people this would not hurt, but for most folks, you for instance, such an accident could decidedly ruin your day. Just a thought!

Also, do not climb towers in the rain and especially during electrical storms. Hey, don't laugh at this one! It has happened, and people have been killed. I mean, I wouldn't think of doing that, and I'm sure you wouldn't either, but there have been HAMs fried by lightening while climbing their towers during a storm.

Another good safety device for towers is a restraining platform that prevents unauthorized ascents of the tower. You know, people (like children) climbing it that you don't want climbing it. These can even be homemade as long as they block the tower at a low level.

These should be constructed in such a way as to be remov-able when you need to go up, but, also so that nobody can get past them unless they are removed. A folding platform, with a sturdy padlock, is a common method of achieving this goal.

One last warning concerning antennas: when installing them and their supports, keep the darn things way away from power lines. I've heard that it is safe if there is 10 feet separating the lines and the antenna. For me, that's not enough. I don't want that antenna anywhere near a power line. Maybe I'm being overly cautious. What-ever the case, try to stay as far away from power lines as possible.

Okay, enough about antennas, bless their little hearts. Let's move on to some safety precautions involving other areas of amateur radio. Yes, there are other areas to discuss.

Most of the rest of this concerns electrical shock in one form or another. For example, don't stick you hand inside an operating radio even if it does use just 12 volts. Many circuits generate high voltage for use with the final output sections, and these can zap you.

While we are on the subject of internal power, be sure that any homebrew equipment is properly shielded, with a case or other enclosure, to prevent accidental shock. It is all too easy to brush against the wrong component with an open frame type arrangement and wind up on the floor, and that's if you are lucky. In all seriousness, some HAM equipment produces voltage/current levels that can be lethal.

Another safety precaution is to ground everything. Remem-ber, electricity is trying to get to ground, and it takes the easiest path available. If you have the equipment grounded, that is the easiest path. If not, you may end up the easiest path, if you are available.

When working with commercially built equipment, please be sure you know what you are doing. Do not poke around inside an operating radio to see what happens. What might happen is that you will acquire a new hair style. Very straight and pointing north! Get the schematic if at all possible, or some help from someone who does know what he/she is doing. That route is a lot less trouble and far less painful.

On a slightly different track, unplug all equipment from both the antenna(s) and 120V AC wall sockets anytime there is serious lightning in your area. I have mentioned this before, but it merits repeating!

Lightning bolts carry enormous power, albeit brief in nature, and if you get hit, your equipment is history. Also, a hit doesn't necessarily have to mean a direct hit. It can come from some distance, as power at that level inducts over a long stretch. Thus, it is best to be safe and disconnect.

Something else to remember is that in a direct hit to your antenna, juice travels right on down the cable. This is hard to deal with, but it is best to have the connector on the cable positioned on a material that is not flammable. More than one fire has resulted from electricity coming out of the connector and hitting something flammable (such as cloth or wood).

Last, let me repeat what has already been said. USE COM-MON SENSE! That is possibly the best advice I can give you, as most safety precautions are just that, common sense. More times than not, if it looks dangerous, it probably is. Let that be a red flag alerting you to BE CAREFUL!

Allow what we have talked about to be your guide, and you can join the ranks of millions of HAMs who have never experienced a safety problem. Those are good ranks to be in!

Special HAM Terminology

Introduction

Like many other hobbies, amateur radio has its own special lingo. Much of this special terminology evolved over the years radio enthusiasts have spent searching the airwaves for other stations. It makes the hobby not only more colorful, but also more efficient.

As with many other radio services, HAMs like to keep their transmissions to as much of a minimum as possible (we aren't as bad as the public service folks, though). The lingo and other shortcuts, like Q-codes, are popular methods to accomplish this task. In addition to courtesy, shorter on-air time serves a couple of pragmatic purposes.

It helps to keep RF radiation levels low and to help keep our transmitters cooler, which prolongs their lives. These radios aren't cats; they don't have nine lives. Anything you can do to make operating conditions less taxing goes a long way toward keeping your cherished equipment running.

I'm getting off track here. Let's get back to the subject at hand, and that is the crazy HAM lingo and other special methods of conveying information. I think you will find this section interesting and useful in your amateur radio work.

The Q-Codes

Here is a list of the more commonly used Q-codes. With these codes as well as abbreviations, the opinion exists that both should be used only with CW. However, today most HAMs accept their application with voice transmissions as well.

In this list, I give you the more customary interpretation of the codes as opposed to the actual wording. If you need the exact wordings, they are readily available in most amateur radio texts and from the ARRL. I found the general nature of the phraseology to be a little vague and sometimes confusing, at least at first. I went with what each code most often refers to when used on the air.

QRA   Station name (call sign).

QRG   Tell me your exact frequency (kHz).

QRH   Is my frequency varying?

QRI   Your tone is ____. How is my tone?
                 (1-good, 2-variable, 3-bad).

QRJ   I can't receive you.

QRK   How do your receive me?
                 (1-bad, 2-poor, 3-fair, 4-good,
                 5-excellent).

QRL   Are you busy? I am busy.

QRM   Malicious interference
             (1-nil, 2-slightly, 3-moderately,
             4-severely, 5-extremely).

QRN   Are you having static problems? I'm
             having static problems.

QRO   Increase power.

QRP   Decrease power or low power.

QRQ   Send faster.

QRS   Send slower.

QRT   Stop sending.

QRU   Do you have anything for me?

QRV   Are you ready? I'm ready.

QRW   Please inform _____ that I'm calling
             (___kHz).

QRX   When is next contact to be?

QRY   What is my turn?

QRZ   You are being called by_____ on
             ___kHz.

QSA   Strength of signal

             (1-scarcely perceptible, 2-weak,
             3-fairly good, 4-good, 5-very good).

QSB	Fading signal.
QSD	Defective key.
QSG	Send message at _____ time.
QSK	If you can hear me, break.
QSL	Acknowledgment of contact (QSO).
QSM	Repeat last message.
QSN	Do you hear me?
QSO	Contact with another station.
QSP	Please relay to _____.
QST CQ ARRL	General call to all HAMs and ARRL members.
QSU	Shall I send on this frequency?
QSW	Send on this frequency (kHz).
QSX	Listen on _____ frequency (kHz).
QSY	Change to _____ frequency (kHz).
QSZ	Send each word/group more than once.
QTA	Cancel message number _____.
QTB	Do you agree with my word count?

QTC	How many messages sent?
QTH	What is your location? My location.
QTR	Correct time.

There they are, the most common meanings of the most commonly used Q-codes. Try them! They save some airtime.

Abbreviations

These you don't hear very often on voice transmissions as they were established to make sending Morse code a quicker process which is how they are most often used.

AA	All after
AB	All before
ABT	About
ADR	Address
AGN	Again
ANT	Antenna
BCI	Broadcast interference
BK	Break
BN	All between
B4	Before
C	Yes
CFM	Confirm
CK	Check

CL	I'm closing my station
CLD/CLG	Called/Calling
CUD	Could
CUL	See you later
CUM	Come
CW	Continuous Wave
DLD/DLVD	Delivered
DX	Distance
FB	Fine business
GA	Go ahead
GB	Good bye
GBA	Give better address
GE	Good evening
GG	Going
GM	Good morning
GN	Good night
GND	Ground
GUD	Good
HI	The telegraph laugh-high
HR	Here/Hear
HV	Have
HW	How
LID	A poor operator
MILS	Milliamps
MSG	Message (prefix)
N	No
ND	Nothing doing
NIL	Nothing for you

NR	Number
NW	Now/Resume contact
OB	Old boy
OM	Old man
OP/OPR	Operator
OT	Old-timer
PSE/PLS	Please
PWR	Power
PX	Press
R	Received as transmitted/Are
RCD	Received
REF	Refer to/Reference
RPT	Repeat
SED	Said
SEZ	Says
SIG	Signal/Signature
SKED	Schedule
SRI	Sorry
SVC	Service
TFC	Traffic
TMW	Tomorrow
TNX	Thanks
TU	Thank you
TVI	Television interference
TXT	Text
UR/URS	Your/You're
VFO	Variable frequency oscillator
VY	Very

WA	Word after
WB	Word before
WD/WDS	Word/Words
WKD/WKG	Worked/Working
WL	Well/Will
WUD	Would
WX	Weather
XMTR	Transmitter
XTAL	Crystal
XYL	Wife
YL	Young lady
73	Best regards
88	Love and kisses

Most of these make sense in terms of what they abbreviate. It is easy to see how a CW operator saves time when sending a message using these codes. For example, it is a lot easier to send BCI than broadcast interference, or VFO rather than variable frequency oscillator.

Also, it must be remembered that with Morse code, each letter (character) has a certain dit/dah pattern, and some letters are more complicated than others. When sending N for NO, your signal is dah, dit (_.) instead of dah, dit, dah, dah, dah (_. _ _ _). It doesn't take a rocket scientist to recognize the time, and energy, saved with this system.

Phonetic Alphabet

With radio transmissions, static, distance between stations, signal strength and a number of other conditions can often hinder the intelligibility of what is being said. To help with this problem, amateur radio adopted the Phonetic alphabet as part of normal operations.

In this scheme, each letter of the English alphabet is represented by a word. Those words were chosen so the letters they represent can not be mistaken for another. It is easy to confuse a B and a D, or a B, a D or an E when hearing them over the air. The phonetics Bravo, Delta and Echo are hard to misunderstand. I don't use the phonetic alphabet as much as I should (I'm working on it, though). However, it really is an excellent way to make sure the information, especially names and call signs, is properly conveyed. Try to use it whenever appropriate or necessary.

A .... Alpha	**J** .... Juliet	**S** .... Sierra
B .... Bravo	**K** .... Kilo	**T** .... Tango
C ....Charlie	**L** ....Lima	**U** .... Uniform
D .... Delta	**M** .... Mike	**V** .... Victor
E .... Echo	**N** ...... November	**W** ....Whiskey
F .... Foxtrot	**O** .... Oscar	**X** .... X-Ray
G ....Golf	**P** .... Papa	**Y** .... Yankee
H .... Hotel	**Q** ....Quebec	**Z** .... Zebra
I .... India	**R** .... Romeo	

You may hear some variations on these from time to time, such as Beta for B or Zed for Z, but this list represents the officially sanctioned phonetic alphabet. Use it in good health!

Some Colorful HAM Lingo

Here is a list of some of the picturesque language heard on the amateur radio bands. There are many more, but these have always caught my attention. Enjoy!

Alligators .... People who don't listen before transmitting. These folks are not particularly popular.

Barefoot .... This term applies to the maximum power a transmitter or transceiver can deliver on its own. That is, without the aid of a power (linear) amplifier. With most HF rigs, this will be about 100 watts.

Buckshot .... This term applies to a station that is over-modulating badly. The result is the station's signal excessively spreads out on the band in a manner that might remind you of the spreading of buckshot fired from a shotgun.

CQ .... A call for any station listening to answer.

Dead .... This refers to a band where no activity is heard. This is often due to bad propagation.

Elmer .... A person responsible for getting others into HAM radio.

Full Quieting .... Refers to a signal that is so strong no background noise can be heard with it.

Gear .... Refers to any and all equipment in your station.

Hi-Hi .... This might be added to an On-Air statement to indicate the statement was a joke. It implies humor!

Homebrew .... Equipment you build yourself. Not usually in reference to kits.

In the Mud .... When a station being received is at, or just above the band's noise level, it is said to be In the mud. Often, the signal is not understandable.

Junkbox .... This is the box, cabinet, or any container that holds all the spare parts. This applies to hobby electronics in general.

Kerchunk .... This is an indicator of the noise a repeater makes when it's activated, but no audio signal is present. Doing this intentionally is considered impolite and is also illegal (transmitting unidentified signals).

Machine .... This is another term for a repeater.

Pirate .... This is a bad guy or one who is operating within the amateur radio bands without a license!

Opening .... As in band opening, this refers to extraordinary propagation conditions that allow for longer range contacts.

Radio .... In HAM radio, every transmitter is considered a radio. Sometimes, even nontransmitters are called radios.

Ragchew .... When HAMs get together and shoot the breeze, this is known as ragchewing. Just a term for conversation.

Reading the Mail ..... When you listen to a conversation being carried on by two or more other Hams, you are "reading the mail". This is a good practice to follow, as a lot of information can be derived about the Hams involved, which will come in handy if you make contact with them.

Rig .... Another term for your radio.

Rock Bound .... Here the reference is to a transmitter that is crystal controlled. Since crystals are made from quartz (a mineral or rock), the radio is said to be rock bound or controlled by a rock.

73 .... A popular greeting or sign-off which means best regards.

Shack .... Regardless of its actual condition, all HAM radio rooms are called shacks or radio shacks. This is an old tradition.

Sign .... When a HAM leaves the air, he or she signs off, or signs. This includes the HAM giving his or her call sign.

Splatter .... Much like "Buckshot", this is excessive bandwidth due to over-modulation of the transmitted signal.

Tennessee .... This is a humorous designation for
Valley Indians   the acronym TVI, which means Television Interference.

Ticket .... No, not for speeding! This is your license!

Time Out .... A term that refers to a repeater discontinuing operation due to a station continuously operating on it for too long. Often, the time-out fault is set for around 10 minutes, and if a single contact lasts longer than that, the repeater simply quits, shuts off, and has to reset. Try to keep your individual transmissions on repeaters as short as possible.

Traffic .... In reference to information that needs to be passed along over the radio. Often when checking into certain networks, you are asked if you have traffic for the net.

Work........................ In addition to being what you do to put bread on the table, this term refers to making contacts on the radio. Working 20 meters, for example, means conversations on the 20-meter band.

YL .... This simply stands for *Young Lady*.

XYL .... This stands for *Ex-Young Lady*. Before you get mad, this is not intended to be an insult and usually refers to a *Married Young Lady*.

There you have it! Some of the terms heard as you roam the amateur radio bands. Again, this is steeped in HAM radio tradition, but helps make your QSOs fun too.

Conclusion

That wraps up this section of the book. I hope it has been informative and useful for you. I also hope it has illustrated the vast array of opportunities available to anyone who becomes involved in this hobby. In the next segment, I provide photographs of some of the HAMs in my area.

Hopefully, they are — or will become — fellow HAMs to you.

~ HAM Radio ~

Photo Gallery

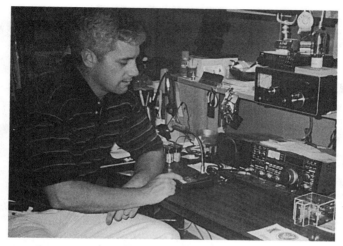

Reuben AD4R has been licensed since 1991. He is now an Extra class HAM.

Jim KR4JY, an Extra class HAM, at his station. Jim incorporates a computer into his shack, using a Hyper log program to keep track of his contacts.

Stuart WD4JRB, in his office at Baptist Hospital, keeps VHF and UHF HTs handy for contacts when he isn't busy.

Mike KB4JHU, an Advanced class HAM, has been licensed since 1984. His interest in amateur radio sprang from a technical school station.

Jim KR4WN is an Extra class licensed in 1992. His interest in HAM radio started with CB radio.

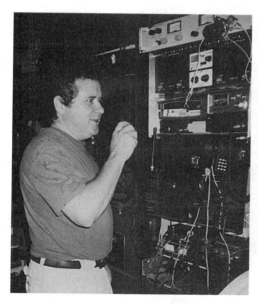

Wiely KE4LTT at his rack mounted station. Wiely, a Technician class HAM, was licensed in 1996. His friends in amateur radio as well as an interest in electronics led him to HAM radio.

Larry Price W4RA, President of the International Amateur Radio Union (IARU) and Frank Butler W4RH, ARRL Southeastern Division Director at the 1999 Huntsville, Alabama, HAMfest

Fred KE4LEX in his closet shack. It folds up so neatly when the doors close that you wouldn't even know it was there. He picked up the hobby when he retired from the Air Force.

Don W3MK, an Extra class licensee, has been a part of amateur radio since 1938! The vintage Collins radio in his workshop has served him well. He also maintains another station in his home.

Carla KG4EVC (Technician class) enjoys keeping in touch with her husband Randy via 2 meters. She hopes to upgrade her license in the near future for those HF privileges.

~ Section 3 ~

On-Air
Operation

Introduction

In the first edition of this book, I covered this area in some detail, but fairly briefly. Since then, I have had several comments directed my way regarding a more extensive discussion of on-air operations. Hence, since there seems to be an interest (and, I have had more time on the air) this section will hopefully expand the topic to the satisfaction, or at least better satisfaction, of those concerned about operational procedures.

As has been already stated with respect to courtesy and proper operating technique, making both part of your personal amateur radio conduct will enhance your experience with the hobby, as well as keep you in good graces with fellow HAMs. Also, it may well keep you out of trouble with the Federal Communications Commission (FCC), as some of the "rudeness" experienced on the air is actually illegal operating conduct.

With that said, let me emphasize that the vast majority of HAMs are very well behaved and follow proper operating procedures. But, as with most things, there are a few "clinkers" in this group. Following the advice given here can help keep you from becoming one, which can often happen without even being conscious of it.

Also, I will discuss some tips (tricks) to working DX, pile-ups, contests, and other related amateur radio communications that will hopefully be helpful. We will further touch on antennas and propagation in more detail, two areas that have significant effect on how well your amateur radio station performs. So, buckle up! Here we go!

Some Operating Procedures and Courtesies

Listening

HAMs have a not so kindly meant nickname for people who commit this first offense. The name is "alligator" and it applies to those folks who DO NOT listen before they transmit! This one is easy to violate, especially in the heat and excitement of contests and scarce DX pileups, but taking a moment to listen to the frequency before you hit the push-to-talk (PTT) button or CW key is a courtesy everyone should develop from the outset.

Often, you will not be able to hear both sides of a conversation (QSO). Hence, the frequency may sound clear when, in reality, another HAM is listening to a station outside your reception range. When you come on the air, you will be "walking on" the other station's signal, and that may interfere with the ongoing QSO. This will not make you overly popular with the other operators! Also, some HAMs are, well, let me just say long-winded, and this can result in silent periods of three, four, fives minutes or longer before the station you can hear responds. By listening for awhile, you probably will pick up on this pattern.

If you think you have found a vacant frequency, one proper way to confirm this is to ask. Sounds simple, but this is often overlooked. Simply go on the air with a message something like: "This is KG4AIC QRZ. Is this frequency occupied?" If it is in use, you will get a quick "yes" or "frequency is in use" to let you know. That basic tactic will keep other operators happy and keep you from looking like an ALLIGATOR!

Another area where listening will help keep down the confusion and flared tempers involves checking into networks (Nets). Here, you will usually be asked to give your call sign, name, and location, and indicate if you have information (traffic) for the net. As might be expected, there normally will be a number of stations checking into the net, and while check-ins are usually broken into segments such as stations with call-sign suffixes beginning with the letters A through E, this can result in a lot of "doubling" (more than one station on the same frequency at the same time). An fundamental way to avoid much of the doubling is to wait for an apparent opening, say just a few words then un-key (release your PTT switch) and LISTEN. You will be amazed how often you will hear another station trying to check in with you when you release the PTT button.

A sample check-in might sound like this: "This is"...un-key. If the air is clear, "This is KG4AIC, Carl, Montgomery County, no traffic for the net and good evening Fred." Fred would, of course, be the operator running the net (net control). In this fashion, you will not interrupt another check-in. Needless to say, this makes things a lot easier for net control.

On a slightly different plane, listening to a station that is working DX (distant exchange) will allow you to get some of the information about that station you will need for your log. For example, his/her call sign, location, and name. Referred to as "reading the mail," it will endear you to DXing stations, as well as other stations trying to work him. In fact, I have heard the DX station operators actually reprimand other stations for asking for such information during the contact. Personally, I feel this is a little "tacky" on the part of the DX station, but many of these guys are trying to make as many contacts in as little time as possible, and they consider having to repeat basic info over and over again an obstacle to that goal. More will be said about DXing, but a general rule of thumb is that the DX station runs the show.

Give Your Call Sign

I can't stress this one enough! Not only is it illegal not to give your call at least every 10 minutes, it is just plain rude. To beat up a little on some of the DX stations, I have heard them go 20 to 30 minutes without identifying. And, sometimes these are the very people who become annoyed if you ask them for their call sign.

Thus, give your call sign frequently! Many amateur operators not only give their own call, but also give the call of the station they are talking to when they "pass the mike" (un-key). For example: "Your signal is really booming in here at 5/9. What is my report? W4AAA from KB6TH (By the way, these are call signs off the top of my head. They probably do belong to someone). In this fashion, and to some degree as a courtesy, the station identifies both sides of the QSO.

Again, *give your call sign when you operate*! It is easy, especially when ragchewing with a friend, to forget to identify. But, it is not good amateur radio practice to do so. So, try to remember this important part of HAM radio. I know one HAM that even has a 10-minute timer to remind him!

Use the Q-Codes

In the previous section, I presented a list of the most commonly used Q-codes. These are not only a part of HAM radio tradition, they are very handy during contacts with other stations (QSOs). Often, especially in the high-frequency (HF) bands, the signals you hear are just above the noise level (QRN) of the band (in the mud), mixed with other signals either on the same frequency or close to it (QRM) and/or involve DXing stations that want to make contacts rapidly. In all cases, employing the Q-codes will make your operating more efficient and pleasant.

Trying to hear a distant signal when propagation isn't at its best can be a frustrating experience, with much of the information fading in and out (QSB). Now see? We are already making good use of the Q-codes. Hence, the shorter the exchange, the more likely you are to perceive the information the other station is trying to convey. Well, that's a long-winded way to say that keeping the QSO short makes operating simpler.

And, this is especially true when working CW (Morse code)! Here, each letter of each word has to be sent individually, so asking "Your QTH?" (location) is far better than having to spell out "What is your location?" Not only is it faster, but much less susceptible to QRN, QRM, and QSB! I know by now you know exactly what I'm talking about!

As stated before, Q-codes are very much a part of amateur radio tradition, but bear in mind that the tradition has its roots in the days when receivers, as well as transmitters, were not as efficient as their modern day cousins. Both selectivity and sensitivity of the receiving equipment have improved tremendously over what they were in the early days of HAM radio, and in those days, short exchanges were almost essential to a complete QSO. So, there was, and still is, a very practical side to the Q codes. Your application of them will be appreciated by your fellow HAMs!

Amateur Radio Jargon

As with many hobbies, pastimes, or businesses, there is a special language (jargon) that comes with HAM radio. These are, of course, terms and/or phrases that have a distinct meaning when applied to said hobby. Again, in the last section I presented a partial list of terms and phrases frequently used in amateur radio. Understanding the meaning of this list will assist you in comprehending on-air conversations. And, that will further enhance your enjoyment of this great hobby!

Naturally, none of this is etched in legal stone. It is all part of HAM radio tradition, but using said jargon will makes things a lot easier for you. It will make QSOs with other stations far more proficient, and it will help establish you, in the eyes of your contemporaries, as a serious amateur radio operator. And, even if the latter is not particularly important to you, it is good for a hobby that has its share of hurdles in terms of popularity and survival.

On a final note, try to use and understand that special language known as amateur radio jargon. I think, in the long run, you will enjoy this part of the hobby once you get used to it!

Band Plans

First, let me explain again what the band plans are and how they work. Up front, let me remind everyone that, while supported by the FCC, these are not a matter of law (regulation), but a matter of choice. I should say choice and cooperation! They have been established in an effort to make the best use of the frequencies we are allotted by the Federal Communications Commission.

They work like this. Each band, be it the 80-meter, 20-meter, 2-meter, 70-centimeter, or other band, is divided into areas of special operating interests. These "interests" might include code operation (CW), phone operation, digital modes, amateur television (ATV), packet, repeaters, and so forth, and the band plans merely set aside certain areas of the band for those operations. In this fashion, CW operators are not interfering with phone stations and vice versa. Additionally, an ATV station will not "walk on" a 440Mhz repeater.

This is a subject of sometimes-intense debate among members of the amateur radio community. There are those who resent the band plans, feeling they are constrictive to overall operational freedom. I will tell you up front, I'm not one of those people! For what it is worth, I feel the band plans are essential to preventing chaos in many of the assigned HAM bands.

Now, I'm not going to make a blanket statement like that without backing up my position. HAMs have what appears to be a good bit of radio spectrum at their disposal, but when compared to the total available spectrum, it "ain't all that much!" And, that dictates that we use the spectrum to its best advantage! The band plans accomplish that!

As previously stated in Section 2, you do not have to follow the plans. That is to say, if you desire to operate phone in the CW portion of 20 meters, there is nothing legal to stop you, provided you are working the portion of 20 meters your license class allows. However, other HAMs are not going to take this kindly, and they may well allege that you are causing "malicious interference" or committing some other offense that could be addressed by the FCC rules and regulations. So, in the end, it is best to adhere to the band plans, as it is not only in the hobby's best interests, but it will also keep you on the good side of your colleagues.

The Phonetic Alphabet

In Section Two, I listed the accepted phonetic alphabet. It is a good idea to learn these. Often, band conditions are such that spelling out your call sign, town, or name in phonetics is the only practical way to convey that information. This, of course, will make the QSO easier for both parties.

As with much of HAM radio, pragmatism gave rise to the use of the phonetic alphabet. In the early days, when phone operation was just coming into its own, this system was adopted as a method of better cutting through the static and other noise on the lower bands. Naturally, I think it goes without saying that the phonetic alphabet is used only with phone communications. It isn't needed with code or digital.

Anne KG4MTI observes her brother Michael KG4KSG, working CW on 30 meters.

And today, while the equipment has gotten better, the high-frequency (HF) bands are just as noisy as ever, and the phonetic alphabet is still as useful as ever. For DXing, it is almost standard procedure to give out your call sign phonetically. A quick sidebar here: The FCC has nothing against using phonetics on the air, but it DOES NOT consider the phonetic version of your call sign as a legal station identification. Thus, at some point within the 10-minute window you must give your call in "regular language" (KG4AIC, as opposed to Kilo Golf Four Alpha India Charlie).

As you become familiar with the phonetic alphabet, you will learn to like it. Too many letters in the English alphabet just sound too similar, especially if band conditions are poor, or you are conversing with a person bearing an unfamiliar accent. In these scenarios, and others, phonetics make perfect sense and will serve you well.

One last point on this subject regards variations of the phonetic alphabet. You will hear many, such as "Italy" for "I" instead of the official "India;" "Colombia" for "C" instead of "Charley;" "Zed" for "Z" instead of "Zebra;" "Honolulu" for "H" instead of "Hotel" and this list could go on and on. Often these variations are associated with geographic areas of the planet. When talking to South America, for example, Colombia, Honduras, Brazil, and other country names are often substituted for the official phonetics. With Europe, England, Italy, Germany, and the like might be employed. These usually make sense when heard, but are not the generally accepted phonetics more customarily heard. A departure from the norm, but one that usually works.

FCC Rules and Regulations

As we all know, in the United States the Federal Communications Commission (FCC) is the controlling authority in terms of use of the airwaves. It sets the rules and regulations concerning virtually all forms of transmitted radiation, including areas such as broadcasting, commercial two-way radio, radar stations, satellite communications, and, of course, amateur radio.

Needless to say, I do not have the room (nor the inclination) to reprint the entire Part 97 of the FCC rules and regulations (the stuff that governs amateur radio) in this text, and that really wouldn't be the purpose of this section anyway. If you want to familiarize yourself with Part 97, and I do recommend it, that segment of the rules and regulations is available from the Federal Printing Office, the Internet, and also from a number of private organizations/ publishers such as the American Radio Relay League (ARRL). Check the source list.

I will prepare you, though. You are in for some heavy reading. The ARRL version runs well over 300 pages. But, it will fill you in on the "Do's and Don'ts" regarding operation on the HAM bands. Also, maintaining a copy of the rules and regulations (from whatever source) in your shack is a highly prudent move. That will allow you to refer to them if an operational question should arise.

Let me cover a few of the rules that will immediately influence your operation in HAM radio. Some of these we have already looked at and some you may already be aware of. First off, you must have an FCC-issued license to operate on any of the amateur radio bands. I realize that may sound simplistic, but I have, believe it or not, encountered individuals who were not aware of this fact. At least, they said they were not aware of that rule. However, since the first section of this book details how to obtain said license, that shouldn't be a problem.

Another rule we have touched on is the "10-minute rule." This simply says that you have to properly identify your station at least every 10 minutes. Interestingly, you don't have to ID right up front. Just be sure 10 minutes doesn't pass before you give your call sign.

Speaking of call signs, these are actually issued to your station, not you. Many HAMs consider their call sign as important, if not more so, than their name. Some even think it is their name. But, in reality, that call belongs to the station.

Transmitting power limits vary acutely worldwide, but in the United States, 1500 watts Peak Envelope Power (PEP) is the maximum amount allowed. Peak envelope power is described, and I quote from the ARRL's *FCC Rule BOOK*, as "average power supplied to the antenna transmission line by a transmitter during one radio frequency cycle at the crest of the modulation envelope taken under normal operating conditions." What this means is that a single cycle of modulation from a transmitter has to stay within the 1500-watt range.

That is in the United States. You will discover, however, that other countries might have different limits. The Australians I talk with tell me they are limited to 400 watts PEP, and some countries — I won't name names — don't seem to have any limit at all considering the strength of some of the signals emerging from their shores.

A few more would include not purposely interfering with a station already on the air, staying within the frequency parameters set forth for your license class and the band you are working, no music on the HAM bands, evaluating your station's radiation pattern as not to endanger the health of anyone nearby (this one applies only to signals over 50 watts in strength) and, of course, no profanity or suggestive language. I feel certain that if you consider these rules, you will see they benefit both HAMs and others using the airwaves.

That is one very good reason to follow the edicts of the FCC, but another very good reason is to keep you in good graces with the law. And, be advised, the FCC enforcement bureau, under one Riley Hollingsworth, is taking violations seriously these days. EXTREMELY SERIOUSLY! As of this writing, there have been a number of amateur radio operators cited by the Commission, and penalties have led to substantial fines, loss of operating privileges, sometimes the offender's license itself, and even visits to the ol' "Stoney Lonesome" (jail).

Now, I don't bring this up to scare you. I remain convinced that 95 percent of all amateur radio operators are law-abiding individuals who have the best interests of the hobby in mind. Just listening to HAMs on the air will lead you to that conclusion. But, as with most things, the few "bad apples" can rot the bottom right out of the barrel! And, it is these people that the FCC is gunning for. With new triangulating equipment at its disposal, the Commission is very well equipped to track down violating stations, and it has been doing just that. Furthermore, I suspect those who have been nabbed aren't particularly happy with the outcome!

Hence, the message here is simply to know and follow the rules. In the long run, you and all of us will be enriched by such conduct, and the HAM bands will be a better and more pleasant place to hang out.

Unauthorized Operation

This fits in well with the last section, and I do want to elaborate on this subject. Sadly, there are people who do not feel they need an FCC license to operate within the amateur radio spectrum. This is seen frequently on the 10-meter band, which is right next door to the old 11-meter band. Eleven meters is more commonly known as the "citizens band" (class D, that is). By any standards, all that can be said about this type of operation in the United States is that IT IS ILLEGAL! For that matter, unlicensed operation is illegal in most countries of the world, but make no mistake about its illegality in this country.

These operators are referred to as "pirates," and they are not welcome in the amateur radio community. If in your travels around the bands, you encounter a station that does not have, or refuses to give a call sign, drop the contact right there! You have probably met a pirate, and further contact with that station is, in itself, illegal. This is not to say that you have to be suspicious of stations that don't immediately identify. Remember the 10-minute rule! Be patient, but if they will not respond to requests for their call, now is the time for paranoia and the time to break off the QSO.

With that said, let me set your mind at ease. The number of pirates you will find, as compared to the legitimate stations, is darn few. Hence, they aren't a big problem. Just be aware that they are out there in the ether and will pollute our atmosphere from time to time. When you do chance upon one, give him or her the proverbial cold shoulder!

HAM Radio Contacts or QSOs

Calling CQ

One of the privileges enjoyed by HAMs is the license (excuse the pun) to get on the air and call for any station that might be within range of their signal. I might be mistaken, but I know of no other radio service that allows this. For example, commercial two-way radio, and that includes the class D citizens band, prohibits this type of operation. So, we as HAMs have the opportunity to randomly call out to fellow HAMs. I guess that is kind of like "reaching out and touching someone." Ehh, maybe not!

In fact, the FCC encourages this, as it keeps the bands active, and to amateur radio, this is known as "calling CQ." Thus, especially on the high-frequency (HF) bands, you are going to hear a legion of stations doing just that — calling "CQ." Naturally, those stations want you to call back (Duhhhh!) I know, I know! I get comments all the time about stating the obvious, but it does keep things clear. Well, at least sometimes!

Okay! There are a bundle of different styles various operators use in calling CQ, and you will have to adopt a style you feel comfortable with. You might try something like this: "This is KG4AIC calling CQ, CQ, CQ. CQ from KG4AIC, central Alabama, CQ, CQ, CQ." I will usually repeat this announcement two to three times before un-keying and listening for responses. That puts the CQ request out long enough for other stations to find and tune into it.

The next step is to listen for replies. There I go again! No, really! You may hear nothing at all, one station calling back, a few stations calling back, or a whole heap of return calls (this is called a "pileup"). Generally speaking, when there is more than one station calling, the strongest stations will get your immediate attention. In this fashion, you will be able to remove those "powerhouses" from the frequency and begin working the weaker stations. That is not meant as an insult to the powerhouses. Not at all! But, it is often best to contact them at the outset, as they will overwhelm the less powerful callers.

And, so it goes. Set the tone for the session (ragchew or quick contacts) and work on through the stations answering your CQ. I realize time is always a factor, but it is considered proper etiquette to work as many stations as the time you have allows, especially when the "quick contact" format is in force. There are few things more aggravating than to wait out a pileup for an hour or more, only to have the station go QRT (leave the air). So, try to accommodate everyone whenever possible.

That is about all there is to calling CQ. There will be occasions when you may have to change frequencies a dozen times to get anyone to come back. But, there will be just as many times when the return traffic is reading "five nine plus 20" on the S-Meter and will be a jumble of almost unintelligible signals. For those times, the challenge will be your ability to pick out a distinct signal to answer. Don't let it scare you, though. The more you do it, the better you'll get at it, and this is one of the foundations of amateur radio! Go ahead! Have some fun! Knock 'em dead! Can you think of any other silly phrases?

Answering CQ Calls

As you are strolling through the band, you happen to hear a signal that sounds something like this: "W6RR, CQ, CQ, CQ. This is W6RR northern California, CQ, CQ, CQ, CQ." Hark! What is this? Well, it is a station just coming on the air, or probably just coming on the air looking for another station to talk with. This is, of course, a station CQing. Frequently, that announcement will be repeated several times before the station actually clears the air for responses, so be patient. Once he or she has un-keyed, you are invited to give your call sign, and if you are heard, your call will probably be acknowledged.

Don't get upset (or mad or sad) if the CQ station doesn't immediately recognize your station. That doesn't necessarily mean he or she is ignoring you. More likely, the station is getting calls from a large number of other stations, many of which you cannot hear. Also, the stronger stations will often get the CQ station's attention first. Better propagation, more power, a better antenna system are just a few of the possible reasons that the other station's signal is stronger than yours.

If you come upon a station that is already involved in CQing, LISTEN to what is going on during contacts with other stations. This is called "reading the mail" and it gives you an idea of the intent of the CQ station. That is to say, is he/she looking for longer conversations (ragchews), or do they want a lot of quick contacts. That way, when your turn comes, you will know how to conduct the QSO. Along these lines, it is polite to respect the wishes of the CQing station. Additionally, the CQ station might use the Q-code "QRZ" (sometimes expressed as "QRZed") to indicate it is ready for the next station to call. That might sound like: "This is W6RR QRZed."

For the short contacts, the station will usually want your call sign, name, location, and a signal report. The signal report would be how well you are receiving the station and is normally given as a numerical entity such as "5/9," "fifty-nine," "5 by 9," et cetera. However, if the CQ station is quite strong, its signal might exceed "S9" and would be expressed as "five nine plus," "20 over S9," "five nine plus 20, or 30, or 40" depending on the signal strength (S) meter reading.

This report gives the station a relative idea of how well its signal is "getting out." I say relative because each radio is different, and these "S-Meter" readings are just ballpark in nature. If your radio is properly calibrated, the "S" reading should be a pretty good gauge for the other station. But, it is hard to maintain calibration, as time and use tend to degrade individual components in your rig. This is unavoidable (sort of like getting old), and will have an effect on the accuracy of the S-Meter readings. Be that as it may, most stations do like to receive some sort of signal strength report even during short QSOs.

For the longer contacts, or "ragchews," a variety of different subjects may be discussed. By the way, all of this applies to both CW and phone contacts, although the CW QSOs will tend to be a little briefer due to the time it takes to send the Morse code characters. My good friend and mentor, Fred K8AJX, tells me that the "more or less accepted" chronology of information concerning QSOs looks like this. During the first exchange (between two stations), call signs, names, QTHs (locations), and signal report (RST) are given. The next information would include a local weather report and the equipment being used. The third exchange would cover the HAM's ages and how long they had been in the hobby. And, if you get into a real ragchew, subjects like each other's profession, other hobbies, how many wives/husbands they have killed, et cetera, would be discussed. Now, now, now! Just kidding about the spousal disposal! You people don't understand me! At least, that's what I've been told!

So, that should give you an idea of what the CQ station wants to hear when you contact it. Again, listening to other QSOs will be a good guide as to the motives of the CQing station. Answering a CQ is one instance where interfering with another station's signal can't be helped. It is acceptable here only because of the nature of these contacts, but try not to abuse it. You will do better by listening for holes in the replies. Of course, there may well be a number of other stations you can't hear doing the same thing, but calling when the "roar" has quieted down just might be the ticket to getting you in. It has worked for me!

Contesting

Here is an area that has a big following, especially among newer HAMs, but also suffers from very little interest amid an equally large portion of the HAM community. I guess what I'm trying to say is that you either love it or hate it. Well, actually, that is a little too harsh. I know many HAMs that do enjoy getting involved in contesting now and again. They just don't live for it.

Anyway, contesting involves a set period of time in which the participants try to make as many contacts as possible under a given set of rules. This may sound easy on the surface, but it can be a tricky and often difficult task. This is primarily due to the fact that there are a WHOLE BUNCH of people trying to do the same thing at the same time. You guessed it! PILEUPS! Pileups, big time!

However, those pileups move quickly, as you will rarely give anything more than a signal report and your location with a good many of these contests. Hence, a CQ station can work in excess of 100 stations an hour during peak times. If the contest runs for 24 hours, as many do, that could mean 2,400 contacts for the station. Scores like that are seen, but it is usually the result of more than one transmitter being used. Maybe one part of the station is sending CW, while another is on phone. The results of both are then combined for the final score. Even if he or she could hold up it would be unusual for a single operator to rack up that kind of score because not all of the 24-hour period is going to be peak time.

So, give contesting a trial run. What is it they say? "Try it, you'll like it!" At first, you will probably find that answering CQ stations the best way to go. You won't make as many contacts, but this approach allows you to get the feel of contesting without the pressure. You can work at a speed that you feel comfortable with, and that makes breaking into contesting a less stressful experience. Just about every weekend, there is a contest going on somewhere in the HF region, and these are great for the beginner.

I have had several people comment that they are hesitant to jump into these contests for fear they will foul up, or foul up the contest. DON'T BE! First, it is hard to foul up a contest single-handedly. Believe me, if I can't do it, neither can you! But, more importantly, you're not going to foul up either! Remember that the stations calling CQ are looking for as many contacts as possible, so they welcome your

Randy KV4AC and wife Carla KG4EVC use a laptop computer during a contest. Both enjoy their HAM radio privileges.

return signal. Here again, just LISTEN to what is going on for a while, and you will quickly pick up on the procedure and information you need to report. Additionally, most of the amateur radio magazines preview the major upcoming contests and will cover the rules and the report the stations are looking for.

If you belong to an amateur radio club, it is likely the club will participate in "Field Day." This is an annual ARRL-sponsored event that is sometimes considered the "Granddaddy" of all contests. In addition to making as many contacts as humanly possible, you will probably get involved in setting up the station in a remote location. That "remote location" can be anything from a tent in the woods to an air-conditioned mobile home, and the power is usually derived from gasoline generators. Some clubs even set up in malls, which is really rough duty!

While this might at first sound frivolous, or even ridiculous, there is a method to the madness (above the contesting aspect). That purpose is to become familiar with setting up and operating remote stations in emergency situations. HAM radio is more formally called the Amateur Radio Service, and the Service part is important! Serving our communities is another HAM tradition, but it also gives amateur radio credibility. That credibility has helped preserve many of our bands. Unfortunately, there are host of profit-oriented types out there that would like nothing more than to strip us of many of our allocated frequencies. I mean, how has the world survived without another paging service? However, because the people who represent the interests of amateur radio are able to point out we perform a valuable service to society, it is difficult for the powers that be to support doing away with us (in effect, reallocating our frequencies). So, Field Day is one of many ways we demonstrate our service and that makes it a worthwhile endeavor.

Okay, enough of that heavy sh....uh, stuff! If you have access to a club that participates in Field Day, I would highly recommend you get involved. I wouldn't necessarily suggest you make this your very first attempt at contesting (considering the number of contests going on, you won't have to), and you don't have to throw yourself into it heart and soul. Just drop by the site and do what you can to help out. Trust me, you will learn a lot, and have some fun!

Let's take a look at the mechanics of contesting. This is much like DXing except that there are going to be "mucho" stations on the air at one time. In some respects, this is good! In others, it's not so hot! But again, don't let me scare you off, as there will be plenty of stations to choose from.

As with DXing, you can go about this in one of two ways. That is, either calling CQ or scanning the band for other stations that are calling CQ. With the high volume of traffic, calling CQ presents the slight problem of finding a clear frequency. Depending on the popularity of the contest, it may take a while to locate said clear frequency. Naturally, it is a good idea to check any spot you think is vacant. You know, "This is KG4AIC QRZ. Is this frequency occupied?" If you hear nothing in return, you're good to go.

The second choice is to scan the band for active stations. This will usually require repeated attempts to make the contact, but as said before, contesting stations work quickly. A typical exchange would sound something like this. "K8XXX QRZ." I would return with "KG4AIC." The other station: "KG4AIC, you're 5/9 from West Virginia, QSL?" "Roger, you're 5/9 Alabama." Many contests are just that simple. In fact, the CQ station doesn't want the contact to last any longer than that. Remember, it is looking for as many contacts as possible, hence, you normally won't have to wait too long.

Also, many contests — in fact most contests — are being conducted on more than one band. Often, the 80-, 40-, 20-, 15-, and 10-meter bands are involved, so this provides additional contacts with the same station, as well as new areas to look. Additionally, some contests will allow you to contact the same station once by CW and once by phone on a given band. This, of course, furnishes even more opportunity to run up your score.

And, that gives you a quick look at contesting. There are so many different events going on in any given year, that it is hard to generalize this subject. Checking out the contest for rules and desired reports before getting involved is probably the best way to ensure you don't step on any toes. They are fun, so try a couple out!

Working Pileups

We have touched on this, but it does deserve a little more discussion. The Pileup is well known to anyone who has been involved in working contests or DX. They can occur at the strangest times and be absent when you would most expect them. Like propagation, they come and go and are hard to predict. I have often wondered why a station in Ohio has a pileup, while just 50 kilohertz down the band, a guy in Afghanistan can't get a nibble! I mean, there is nothing wrong with Ohio, but it isn't the DX contact Afghanistan is. It just works out that way sometimes!

However, if you work DX and/or contests, or plan to work them, you will run into a pileup or two. Actually, they are not as bad as they might sound, especially if you have a good operator on the other end. There are several ways the CQ station can handle them, and experienced HAMs are well familiar with these techniques.

As with contesting, one good approach is to keep the individual contacts short. This will not only increase the number of contacts, but also keep the waiting station from losing interest and moving on. It is immensely frustrating to try and work a station that has a huge pileup but insists on spending five minutes with each contact. That type of behavior will run off many of the waiting stations in a hurry.

Another good approach is to take the calling stations in some form of order. For example, "This is KG4AIC QRZ. Only stations in the number 4 area please." In this fashion, the CQ station can start with "0" and work up, or start with "9" and work down. This technique will vastly narrow the number of stations calling at a given time, which allows far more stations to "get in." Again, it helps discourage despair on the part of the waiting stations, as when your number comes around, you will probably get the contact.

Occasionally, you will encounter a CQ station that makes a list of call signs then goes back and contacts them one at a time. Here, you still have to cut through the pileup, but since the CQ station is listening for as many calls as it can discern, your chances of getting on the list are greatly improved. Depending on a number of factors, this can work very well, but can also get away from the CQ station if it is not careful. However, it is another attempt to deal with the proverbial pileup!

Working a "split" is a method some stations use to manage a pileup, although I have never really seen the wisdom in this approach. Here, the CQ station will transmit on one frequency and listen on another (sometimes several others). You will usually hear something like this: "This is I3HJT QRZ, listening 5 up." What this means is that the station is receiving 5 kilohertz up from where he/she is transmitting. For example, if you are receiving I3HJT on 14.250 megahertz, you will want to transmit back on 14.255 megahertz.

Most modern HF transceivers allow you to put one frequency in one variable frequency oscillator (VFO) and another frequency in a second VFO and by activating the "split" function (usually pressing a button), the rig will listen on the first frequency and transmit on the second. Thus, working a split is not all that hard, but I have found that the pileup on the second frequency is just as bad as if you were responding on the original frequency. Does that make sense? I hope so. I guess if you listen on more than one frequency — This is I3HJT QRZ, listening 5 and 10 up — for example, you will tend to spread the pileup out a little. Anyway, be aware of splits, as they are used and can be a little confusing at times!

So, there are some things to think about when working a pileup. Actually, it is a challenge, and there is a certain feeling of satisfaction when you break through all those other stations and get the contact. It's all part of HAM radio!

Logs, Logging, and QSL Cards

From time to time I hear grumbles about logs and logging, but for me at least, this is an enjoyable part of the hobby. As seen in Section 2, a typical log is a page, or book of pages, that allows you to record the QSO. Such things as the date, times of the contact (on/off), frequency, call sign of the station contacted, mode used (CW, USB, etc.), and comments about the QSO are included. I personally like having that record.

There are times when trying to keep the log up to date, especially during contests, can be hectic, but without that log, you would be hard pressed to remember all the stations you have encountered. Maybe 20 years down the trail, when I have logged my 1 millionth station, I will share some of the less than favorable feelings about logging. However, I kind of doubt it! I least, I hope not!

In addition to paper logs, there are a variety of computer logging programs (electronic logs) that can be downloaded or purchased. That, of course, presumes that you have a computer, which many HAMs do have. I have one such logging program, and to be perfectly honest, I don't use it all that much. I find the paper log to be more than adequate, but the computer programs do offer some features that are nice.

For example, if in doubt, you can search the log for the call sign to see if it has been contacted before. And, if so, when. Also, many programs allow you to "link" with a "call sign site" to get information about the station. This involves the program checking a relevant database for that call sign. Normally, if found, such info as the operator's name, address, license class, et cetera, will be provided. All this is handy!

Speaking of electronic logs, there are companies that design special logs for just about all the major contests; often these logs have to be used if you plan to submit your result in an electronic format. Here is one area where I really do appreciate computer logging. In the heat of activity during, let's say Field Day, these programs are highly useful. They will tell you if the station you are "talking to" has already been contacted (a "Dup" or duplication), and they make logging the huge number of contacts not only accurate but very simple. For a busy contest, this is really the only way to go.

There is a dark side, however! In most cases, the program is putting the data on the computer's hard drive. A good place to put it, as the hard drive has lots of space, but they have been know to fail (crash). If that happens, you will lose all that precious data, unless you are willing to pay a retrieval service big bucks to get it back for you! Hence, it is wise to back up your log with a floppy disk, or some other type of recording media.

So, logging, like so many other aspects, has its place in amateur radio. If you think it is bad now, there was a time, many moons ago, when you had to record every on-air operation, even if it was a futile attempt to get through a pileup. Those days are gone, but the practice of logging contacts lives on!

Next, let me say a few words (You know me by now! Probably quite a few!) about contact confirmation, better known as QSL cards. It has been said a QSL is the final courtesy of a successful QSO, and to some extent I have to agree. I don't send QSL cards to every station I contact, but if a station sends me one, or asks for a card, it will get one back.

Not only is this a courtesy to a fellow HAM, but many amateur operators enjoy collecting the cards. Others are involved in receiving awards for the number of contacts made. These awards come in many flavors, such as contacting 100 countries, contacting all U.S. states, all U.S. counties, and so forth, but they all have one thing in common. You must be able to prove you actually made the declared contacts. You guessed it! QSL cards are, in almost every case, considered proof! So, your cooperation regarding sending QSL cards will be immensely appreciated by those HAMs trying to earn awards.

Often called "wallpaper," they could actually be used for just that purpose. In fact, I have visited HAM shacks where the walls were virtually covered with QSLs and other related documents. Hence, if you are looking for some nice wallpaper for your shack, keep QSLs in mind!

One caution here! When sending or returning a card, be sure to check the information against your log for correctness. Again, this is most important for those who are interested in using that card for award confirmation. Any QSL that has incorrect information will be disqualified as proof for most awards. Also, if you foul up, start over with a new card! QSLs with white-out or other changes will not be accepted, either. Thus, keep the card clean and accurate, as it is only useful to a fellow HAM if it is in that condition!

Conclusion

Well, I hope I have provided some useful information regarding actual on-air operation. If you are like most of us (HAMs), you love the hobby and all that it furnishes. In that vein, we want to protect amateur radio by operating in the most appropriate fashion possible. In the spirit of "If you're going to do, do it right," there is a certain pride associated with proper conduct on the bands. But, I know I don't have to impress that upon you!

I'm sure I have missed a few things in this discussion, and as you progress in your pursuit of HAM radio, you will encounter additional proper operating methods and skills. I know that is how it has worked for me. Amateur radio is a friendly and fascinating world, and the conduct you display when on the air will noticeably shape your fellow HAMs' opinion of you. But, with the knowledge in this section at hand, you will instill favorable impressions on everyone you meet on the bands!

Working with the Equipment

Tuning the Radio

That is not intended to insult you! In this section, I merely want to discuss some of the finer points regarding using HAM radio equipment. And, that is not always as simple as it might seem. Most of the gear on the market today is great, but it can be a little tricky from time to time. I will never forget the first time I got my hands on a "for real" single sideband transceiver. I was like the proverbial kid in a candy store, but it took me some time to get a handle on tuning that rascal.

I'll guarantee you it wasn't the equipment. It was me! When in comes to jumping in and getting our feet wet, some of us HAMs tend to be our own worst enemy, or in the words of.......someone, "We have met the enemy, and it is us." Often, however, we get more than just our feet wet. So, for this, and other reasons, let me give you the benefit of my experience (folly) in terms of operating amateur radio equipment.

Alright, back to tuning. There is an art to this, as many of the HAM bands can be crowded, and with some modes, tuning is very critical. This is especially true of the high-frequency (HF) arena, where single sideband (SSB) is often used for phone communications. Unlike amplitude modulation (AM) and frequency modulation (FM), SSB signals take a little more care to tune in properly. With a sideband signal, the transmitter has filtered out the signal carrier and one side of the modulation cycle. Thus, all that is left is the other side of the modulation cycle; either "upper side-

This is a typical 1980s vintage high-frequency (HF) transceiver. This one covers all bands from 160 to 10 meters and employs upper and lower sideband as well as CW and RTTY. The highest power this unit will produce is 100 watts.

band" or "lower sideband" depending on which side was filtered.

As might be expected, this tends to narrow the bandwidth just a tad! Actually, it narrows it to about one-third of analogous AM or FM signals. And, for that reason, you have to be delicate in your tuning technique. Here, a few hundred kilohertz can be the difference between an intelligible signal, and one that sounds like Betty Boop and/or Donald Duck. Am I showing my age there, or what?

Anyway, with sideband I usually try to go from the high-pitched signal to the lower pitch. With upper sideband, that will be from a lower frequency to a higher frequency and with lower sideband, the reverse. Eventually though, the signal will become so low in pitch that it is no longer understandable, hence the properly tuned signal is somewhere in between those two. At some point, the voice will sound fairly normal, although a certain amount of "yarg" (the sound characteristic of sideband) will always be present.

Now, that is only my preference, and I know HAMs that like to tune backward (you will notice I employ a term synonymous with WRONG for the other method). What is important here is to find a technique that works for you! Just because my approach is the more commonly accepted procedure, doesn't mean that other way won't work. You'll just do better if you tune from high to low!

Okay! Okay! I need to put my ego on hold for the time being. As I stated, going from a higher pitch to a lower pitch is usually the recommended method. I mean, that didn't come from me! But, you may well find the reverse easier to do. In all honesty, I often use both directions. The important part is that you get the station tuned in properly as when you answer him or her you want to be on the same frequency. One of the mistakes I made with an early rig was to use the "Receive Increment Tuning" (RIT) control to "clarify" the signal. This is a fine-tune adjustment that allows you to leave the transmitter on a specific frequency, then slightly retune the receiver to clear up the tone and pitch of the other station's signal. This worked fine for the sound at my end, but when I transmitted back, I was off frequency, and often asked to retune my station. It took a while for me to figure out what I was doing wrong.

Another problem that will make tuning the SSB signal hard is band traffic. During peak periods, there are going to be spots on the band where several stations are right on top of one another and separating them is a challenge. In this scenario, you will hear high/low pitch signals that seem to blend together. The best (in fact, only) way I have found to deal with this situation is to try to single out the strongest signal. In that fashion, you can often at least hear well enough to tune that station in. Some assistance can be sought from the filters built into many rigs. These will allow you to better isolate an individual signal among a block of overlapping stations. Just remember to sneak up on those signals. Due to the bandwidth, you can easily overshoot a station by moving too quickly.

Along these same lines, vigilant tuning also applies to CW signals and the digital modes (RTTY, AMTOR, PACTOR, PSK-31, etc.). In each case, bandwidth is narrow and for many of the same reasons expressed above, careful tuning is essential to good reception. With CW, it is best to listen for the tone, or pitch, of the signal, and tune accordingly. Most amateur operators send code at a tone of between 700 and 900 hertz, so this can be of aid in "zeroing in" on the correct frequency. Again, the proficiency of your reply will depend largely on your signal being on the right frequency.

With the digital modes, bandwidth is even more confined, which leads to even more delicate tuning. Fortunately, many of these modes use computer programs, or dedicated terminals as part of the reception/interpretation process, and these have tuning aids built in. That is probably a good thing, as some of these signals would be extremely difficult to pin down if not for this tuning assistance.

So, that provides a few tips that hopefully will serve you well. Ultimately, the proficiency of your contacts will depend largely on how well you tune your rig. The better the other stations hear you, the better the chances are they will answer your call or CQ. But, fear not! With a little practice, you will be tuning like an old-timer in no time at all.

Tuning the Antenna

Now that we've talked about tuning in other stations, let's talk about the antenna you're using to pull those stations in. With respect to tuning the antenna, we get into some gray area, as this subject covers several different aspects. There is the question of antenna design/construction and if the antenna is "resonant." There is the topic of the so-called "antenna tuners" and there are antennas that utilize traps and/or special capacitance devices that "tune" them to various bands.

In each case, the idea is to provide an apparatus that radiates in the most efficient fashion possible. One fact about amateur radio that everyone learns quickly is that power is not the most important thing! Antennas, and our next subject, propagation, play a far more pivotal role regarding the effectiveness of your station. Hence, we want to have the best antenna systems we can build and/or afford, and sometimes that means using more than one of the above tuning methods.

One of the key factors involving antennas is the subject of "resonance." Each individual frequency has an equally individual wavelength, and the trick is to design the antenna to have the highest radiation at that wavelength. This is called resonance! Of course, it would be nearly impossible, or at least highly inconvenient, to have a separate antenna for every frequency, so we fall back on the next best thing — antennas designed to be resonant on individual amateur radio bands, or in the case of many antennas, several bands. This, like so many things in electronics, is a compromise, but manufacturers are very adept at making this as good a compromise as possible.

When I first earned my Technician Plus license, I needed an antenna to operate on the 10-meter band. I considered a number of options and finally settled on a dipole antenna tuned to 10 meters. That is, its length was cut so that the antenna performed best in the 28 to 29.7 megahertz wavelengths. Now that is a lot of spectrum, and few antennas can actually cover all of it. So, another compromise came into play. Since I could only use the 28.1 to 28.5 megahertz portion of 10 meters, the antenna length was selected to suit that end of the band. Now I had an antenna that was resonant, or tuned, to the frequencies I would be operating on.

Okay, that is one way to go about it! I could also have just as easily selected a multiband dipole, vertical or beam, or a single-band vertical or beam (a "monobander"). However, at the time, the single-band dipole seemed a good choice. As it turned out, a friend gave me a "tri-band" vertical when I received my General ticket, and that expanded my capability to include the 15- and 20-meter bands. But, what is important here is that in each case, the antenna system has to resonate at the required frequencies (bands).

While we are on the subject of the tri-band vertical, let me describe how it works. This antenna utilizes two methods to accomplish the goal of resonance on a particular band. First, the system employs "traps" in its design, and as the name might imply, these coil, or coil-capacitor mini-circuits prevent, or trap, the frequencies you do not want to radiate from the antenna. Since the antenna is physically designed to handle three different bands, without this technique frequencies from all three would emerge from the antenna. Traps are also frequently employed in "Yagi" beam antennas. The second tuning process involves a box in the shack that electrically adjusts a variable capacitor at the antenna. Here, the idea is to match the impedance of the antenna with the impedance of the radio (usually 50 ohms). So, when I want to work 20 meters, I hit a switch on the box that activates a motor that rotates the variable capacitor. This changes the capacitor's value, which, in turn, adjusts the impedance. A meter on the box indicates which band the antenna is being tuned to. When it gets to 20 meters, I stop tuning. Now that I'm in the 20-meter "ballpark," I activate the transmitter and further adjust the box to get the standing wave ratio (SWR) down to its lowest point. This is done using the transceiver's built in standing wave ratio (SWR) meter. When that is accomplished, the antenna is properly "tuned" for 20 meters.

By the way, transmitters can be adjusted using a "dummy load," but antennas cannot. Thus, you will have to actually go on the air to adjust this type of antenna system. Naturally, you will want to tune the antenna on a clear frequency, and since it is an on-air operation, it is necessary to identify your station either before or after the tuning process. Remember, the 10-minute rule!

Utilizing these two functions, this antenna system can be brought to resonance just about anywhere within the 10-, 15-, or 20-meter amateur radio bands. And, while vertical antennas are not often thought of as optimum performers, this one has done a remarkable job for me!

Next, let's look at the "antenna tuners", otherwise, and probably more accurately, known as "transmatches." I say that because, in reality, an antenna tuner does not actually tune the antenna. Instead, it matches the impedance of the antenna to the impedance of the radio. Thus, the "trans" for transmitter and "match" for the matching operation. In doing so, the antenna tuner ensures the lowest possible SWR, which translates into good signal radiation.

As might be expected, the tuner is connected between the transmitter and the antenna. Most of these devices employ either variable coils (rolling inductors) or variable capacitors, some both, to perform their magic, and are a wise addition to any HAM shack. A really good tuner can literally match the rain gutters on your house to your radio, although, I don't necessarily recommend that. I think you will find that a more conventional antenna will do a far better job!

So, that covers some of the finer points concerning antenna tuning. This has really only scratched the surface, however, so I do recommend you do a little research on a specific antenna you might have in mind. Again, the two important points are resonance and SWR. If both of those are "cookin'," you will be in great shape!

Antenna Installation Tips

What I plan to present here is a mélange of tricks, secrets, and inside information that I have picked up along the way from other HAMs and personal experience. As such, it is intended as an aid to antenna placement, but does not profess to be inclusive by any means. Conversations with other amateur radio operators, especially the experienced ones, will also garner useful information on this subject. With that said, let's start with where to put an antenna.

Now, this at first glance might sound rather rudimentary, but there is more to placing an antenna than you might think. First, you need to consider safety! By all means, keep ALL antennas away from power lines. The rule of thumb is at least 10 feet away, but personally, I like to keep them as far away as possible. Not only is there a very real risk of electrical shock involved, but those power lines can interact with your antenna. And, that can result in the radiator not performing up to par! Bottom line: power lines are a big NO-NO!

Another consideration regarding antenna location and safety involves accessibility. You want to place the antenna system in an area that will keep unwanted visitors away, especially the towers that are made to be climbed. Too many times in the past children and adults alike have been injured, or worse, because they had unnecessary access to an antenna. So, don't let that happen!

The height of the antenna is another factor that must be considered. Remember that you will be transmitting a signal from that antenna, and close proximity to that signal can be hazardous to the health of people and animals. How close is a dangerous proximity? That depends on several factors, such as the type of antenna and the power being applied to it. The FCC has a whole section in Part 97 regarding what it refers to as RF Exposure Limits. When you get your copy of the FCC rules and regulations, I strongly suggest you take a look at this section. It will guide you in determining if an antenna location you have in mind is safe in terms of radio frequency (RF) radiation.

Additionally, omnidirectional antennas do not present as bad a safety hazard and the directional beams. This is because the beam-style radiators concentrate the transmitted power in a narrow pattern. And, as HAMs, we are allowed to use up to 1500 watts of power on most of our bands. Trust me, that is enough to fry just about anything if it is close enough to the antenna. Hence, this becomes an important contemplation when it is time to decide where to put that ol' 20-element Yagi!

On a lighter note, you will also want your antenna location(s) to be convenient to your radio shack, if for no other reason than long runs of feed line (cable) tend to "soak up" some of your signal. So, try to position your mounting pole(s) or tower as close to the shack as possible. In the end, you will thank yourself (and me)!

There is in HAM radio a the-higher-the-better mentality regarding antennas, that makes a lot of sense. The FCC does limit the height of our antenna systems to 200 feet, unless otherwise approved, and that is usually more than enough for most HAMs. But, it is a prudent idea to get your antennas up as high as possible. Now, there can be a few pesky obstacles to this, of which the most prominent would be neighbors, neighborhood restrictions, and neighborhood, city, and/or county antenna covenants. Naturally, we don't want to inconvenience or offend anyone in our communities, especially the people we live around. So, most of us try to keep our antennas as unobtrusive as possible. And, that usually means keeping them considerably lower than the 200-foot maximum. These restrictions are being seen more than ever these days, and it is wise to investigate such covenants whenever you consider renting or buying a new home.

Also, keeping your antennas as aesthetic as possible will go a long way toward keeping the neighbors happy. Large, ugly antennas are known as "skyhooks." In the words of

Harry Helms, AA6FW, a skyhook is "an antenna, especially a large, grotesque one that upsets the neighbors and frightens small children." Now, as good HAMs, we don't want to be frightening small children (at least, not most of them). So, it is in our own best interests to utilize antennas that don't do that.

However, in the spirit of good radiation, we also want to get the antennas up as high as possible. Here we go again with "compromises!" Many of the antennas designed for amateur radio are not all that obstructive in appearance, so if you keep them at a reasonable height they probably won't be noticed by the neighbors. Clusters of trees are a great place to hide an antenna. If practical, try to get the antenna a little bit above the top of the trees, but if that is not possible, older tree stands can afford 60 to 70 feet of altitude. A good friend of mine, Don W3MK, has his tower and Yagi beam almost completely concealed by surrounding trees. You literally have to get into his backyard to see the antenna. Unfortunately, during the winter, when the trees lose their leaves, the antenna system is a little more visible. Even so, this approach beats a 60-foot tower topped by a 30-foot Yagi sticking out of the roof of your house.

In really strict neighborhoods, don't forget the attic! Many a HAM has his or her dipole, vertical, or even Yagi in the attic of the home. While this is not ideal, it can be a way around the "NO EXTERNAL ANTENNAS AT ALL" covenant often found with the newer housing communities. And, especially with HF, you will get out! It might not be quite the same as if you had the 100-foot tower, but it is better than not getting on the air at all.

Alright, enough about the obstacles. Let's get back to installing antennas. Aside from raising the antenna to the highest possible altitude, keeping them away from large metal structures (dense power/telephone lines, metal roofs, etc.) is very beneficial. While large metal surfaces can sometimes act as reflectors, more frequently they merely interfere with the proper operation of the antenna. Thus, it is best to try to avoid close proximity to anything large and metal.

When it comes to performance and stealth, there are few antennas better than the dipole. A dipole is simply two lengths of wire cut for the intended wavelength, stretched out at roughly 180 degrees from each other, with the feed line connected at the center. I say roughly, because there are some variations to this that I will cover in a moment. The drawback is, again, space! At 80 meters, those two wires are going to be around 67 feet each, and that would mean a total length in the neighborhood of 134 feet. That is for a "half-wave" antenna with two "quarter-wave" legs. Now, I surely don't have that kind of space! Not without running a quarter of the antenna out into the street in front of my house!

However, there are some ways around this. One method would be the "folded dipole," where part of that length is "folded" back on the rest of the antenna. This approach will shorten a dipole considerably. Another technique is to close the angle between the two elements or "legs." Instead of the normal 180 degrees, the elements could stretch out at 90 degrees from each other. In this fashion, you can use the corner of your property to install the beast.....uh, beauty!

And, a third solution is to set the dipole up as an "inverted V" or as a "dual sloper." With the inverted V, the center of the antenna, where the feed line is connected, is higher than the ends of the legs. This produces a profile that re-sembles an upside down letter V. This, of course, tightens up the length of the antenna, at least a little. The dual sloper uses legs of different lengths. This arrangement al-lows for badly needed flexibility when space is a consider-ation. Again, it is best to keep the elements 180 degrees apart, but if your property is notably longer than it is wide, this style antenna could be the ticket.

So, you are going to put up an external antenna. Do you use a mast or a tower? The answer to this question will depend on a number of factors. These will include the type of antenna, the height, size, weight, and so forth. As might be expected, towers will sustain larger, heavier antennas than the masts or poles, so that might well be your first consideration. Also, towers are favored for heights much above 40 feet because they are more durable structures. Additionally, if you are installing a beam, it will need a ro-tor, or motor to turn the beam in the direction you want to transmit. Here again, the very nature of the tower's archi-tecture (normally a hollow triangle) lends itself well to mounting that rotor. As a last note, towers come as "free-standing" or "guyed." The freestanding structures are self-supporting and do not need additional assistance, while guyed towers need the aid of guy lines to stabilize the ar-rangement.

On the other side of this coin is the mast, which has its own set of attributes. They are easier to secure, lighter in weight, and much easier to handle. While masts probably won't support large beams, for verticals and dipoles they do the job nicely. Mounting a rotor to a mast is not as convenient as with a tower, but it can be and is done. Thus, a mast is often a good choice for a small beam, such as one for 2 meters.

As far as actually installing the mast, this can be done from the roof of your house, from a chimney, or from the ground. With a ground placement, the mast can be secured at both the ground and at the eve of the building. That is usually all that is required to keep the whole system in place. That is assuming the total height is not going to be much above 40 feet. If you need the antenna higher, guy lines can be incorporated to provide additional support for the mast. This is also a good method to use when mounting the mast on top of a roof. With a chimney arrangement, straps are employed to secure the mast to the chimney. In most cases, they are more than sufficient to keep the mast fastened in place.

In a final weighing of these two methods, two factors have to be considered — wind and other weather. As might be expected, triangular-shaped towers have better resistance to strong winds than the single pole masts. Hence, if you are located in an area where strong winds are frequent or a problem, you might be wise in leaning toward a tower. I have seen masts literally bent in half by powerful wind gusts. However, if wind is not a real threat, then a mast might well do the job. Remember that wind also puts pressure on the antenna itself (wind resistance), so take that into consideration when evaluating your wind situation.

To a lesser extent, the weather, especially thunderstorms, plays a minor role in which type of support you choose. In areas of the country where storms are prevalent, lightning becomes significant to the equation. Lightning just loves metal, especially metal that is sunk in the ground! And, since a tower contains far more metal than a mast, it can be a better target for those bolts from the blue. In other word, towers make better lightning rods!

In all fairness, though, this aspect is probably not as pertinent as it might seem. Lightning is strange stuff, with a mind of its own, and I have seen it pass up towers and other metal structures to strike an open field. You just don't know what it is going to do. So, as stated, this factor is less crucial than others.

Conclusion

Well, that became a tad more long-winded than I intended, but this subject is well worth the effort! I can't say enough about the importance of your antenna. This factor, as much as (probably more than) anything else, will affect the efficiency of your station. The 1500-watt linear amplifiers are nice additions to a shack, but they will not influence the range and clarity of your signal nearly as much as a good antenna system. And, you don't have to take my word for it! Ask any experienced HAM, and you will hear the same tale!

I hope this section has provided information you will find useful when installing antennas. This is a topic that could easily be a book in itself, but I have tried to cover the areas that have been significant to me, as well as other HAMs I have discussed the subject with. Again, there is a profound wisdom is talking with and LISTENING to the old-timers. They possess a wealth of knowledge concerning all aspects of amateur radio. Whenever I have questions, I touch base with the experienced HAMs in my area; when it comes to antennas, it is hard to do justice to their merits and contributions!

Propagation

Explanation

In this section, we are going to take a look at a topic that has substantial dominance concerning the performance of your station — propagation. It can be a very mysterious subject at times! Hopefully, I will be able to help you better understand at least some of the key factors involved in this area without totally confusing you and me!

Okay. What is this enigmatic phenomenon? To understand propagation, we must have some understanding of the atmosphere that surrounds the Earth. This is what creates the signal behavior associated with propagation. The area of most concern to us is the span from 40 to 400 miles above the Earth's surface known as the ionosphere. The ionosphere lies above the stratosphere about (6 to 25 miles up), which lies above the troposphere (from the surface to roughly 6 miles). Within the ionosphere are several individual layers that are affected by solar energy. Changes within these layers are responsible for what happens to our signals.

With all that said, let's look at the individual layers and how they are affected by the sun.

The layer closest to the Earth is the "D" layer. It runs from about 25 to 55 miles up (actually, the area of 40 to 55 miles will have the most influence.). The D layer is virtually a daylight layer that disappears at night. Thus, bands such as 10 meters that depend on the D for long-distance communications are in trouble after the sun goes down. There are exceptions to this rule. I have seen 10 meters open late into the night, but that is rare!

The next layer in line is the "E" layer. Here, the range is between 55 and 90 miles above Earth. The E layer is also a daylight layer, although it lasts longer after sunset and it displays some unusual characteristics at times. One is the tendency to become patchy. These patches have been known to propagate both high and low frequencies, including VHF. From time to time, you will experience some rather long-range communications on 2 meters (sometimes a couple hundred miles or more); it is the E layer that is largely responsible for this anomaly.

The highest and last layer is the — you guessed it — the "F" layer (if you have been paying attention, you have noticed that these layer designations are sequential, alphabetically speaking). During daylight hours, the F layer is actually two layers, F1 and F2. The F1 layer extends from 90 to 250 miles up, while F2 is anything above 250 miles. It is these atmospheric bands that provide many of the really long-distance QSOs.

At night, the two F layers combine into a single layer, generally considered to be around 180 miles above the Earth. This is due to the diminished solar energy. That causes the two layers to merge into one. Since the F layer is normally the only one still in existence after dark, all those distant stations you talk to at night come in off the F layer. Hence, we don't want anything to happen to the F layer, or any of them for that matter, as without the ionospheric bands, there would be no DXing.

Why is that? Well, that's a very good question! One that I will attempt to answer. Solar energy — ultraviolet radiation from the Sun — causes these layers to ionize. Ionization is the process by which extra solar energy causes atoms within the atmosphere to absorb additional electrons. This gives the atmosphere a negative charge, and since your signal is negative, like two permanent magnets with similar poles placed end to end, the signal is repelled or reflected by the atmosphere. This electron absorption results in the layers of the ionosphere bouncing your signal back to Earth — sometimes at great distances — instead of allowing it to pass through and into outer space (never to be heard from again!!!).

Of course, that is what we want! When I send my signal out, I don't want it to end at Mars! I want it to be reflected back to Earth, say around Japan or Australia. However, that doesn't always happen. It might come back down in Europe or South America, but that's okay! I like talking to those folks, too! The main concern here is that it doesn't just disappear over the horizon and out into the ether. Naturally, I'm referring to HF signals, as VHF and UHF signals, unless acted upon by some other force, do just that — disappear into space.

Now, you might be wondering about this solar energy stuff, and that brings us to a short description of what is known as the solar cycle. Solar cycles involve the amount of solar activity on the sun's surface and have up and down segments that blend into each other. A full cycle lasts on average 11 years, but they have been known to be shorter than that. The high, or up segment (peak), of the cycle is characterized by increased sunspot and other solar activity. Regarding most HAM bands, this activity produces better propagation. I say most, as the lower frequencies can actually be hindered by this activity. This encompasses a peculiarity of these frequencies to be absorbed by the ionized layers, as opposed to being reflected by them.

The low, or down half, of the cycle (minimum) is often associated with very poor propagation, except for the lower frequency bands. Here, a decrease in sunspot numbers results in less ionization, with the consequence being less signal reflection. However, the 160- and 80-meter bands actually benefit from this lack of solar activity.

Each solar cycle is an entity all its own; hence, no two are exactly alike. The most recent cycle would be defined as normal, where as the cycle that occurred in the late 1950s and early 1960s was considered GREAT! One HAM I know was able to make a fairly long-distance contact with a dummy load connected to his radio. He didn't intend to do that, just forgot to switch over to the antenna. But, that very vividly illustrates how good the propagation was during that cycle.

Thus, these solar cycles are extremely important in regard to propagation. In fact, they could almost be considered everything. Fortunately, even during the down side, some communication is possible, so amateur radio doesn't just die when the cycle is low.

To cap this off, I think it is fair to say that the atmospeheric (ionospheric) layers and sunspots are our friends! And, don't forget it! Just kidding, I know you know that! They do our bidding in terms of long-distance communications. But, there is a little more to it than just that, so let me discuss some of the types of propagation you will encounter as a result of the D, E, and F layers and solar activity.

Types of Propagation

Sporadic-E

In addition to normal ionospheric characteristics, that ol' atmosphere displays some surprises from time to time. Among these is Sporadic-E. I'm starting with this one because I have already touched on it. As previously stated, the E layer will occasionally develop cloud-like patches that act as terrific reflectors. This is good news for operators on frequencies in the 10-, 6-, and 2-meter bands, as this is where Sporadic-E will help the most! It has been known to show up on the 1.25-meter band, but that is rare, and that band is generally considered to up the upper limit for this type of propagation. Or, is that the outer limits? Well, it's something like that.

Since most propagation is better during the daylight hours, it comes as no surprise that Sporadic-E is best during the day, usually peaking during the midmorning hours. That is not to say that you won't see it at other times! It can and does appear virtually anytime of the day or night if the conditions are right!

The duration of a Sporadic-E event can be from a few minutes to hours. Again, the prevailing conditions will dictate its longevity. It will tend to last longer on 10 and 6 meters than 2 meters, but that too can change. Since the patches are in motion, it is hard to predict how long or in what direction the propagation will be. Also, the spring months seem to be most favorable for Sporadic-E; however, you may well see it again in the winter months.

Tropospheric Ducting

Our next topic, Tropospheric Ducting, also aids the VHF/UHF operator. As we discussed earlier, the troposphere is the closest atmospheric band to the Earth, and it is here that the planet's weather is born and thrives. As a rule of thumb, the higher up you go, the colder it gets (usually about 3½ degrees Fahrenheit for every 1,000 feet of altitude). However, as a result of large weather fronts, or the sun warming the upper atmosphere sooner than the lower regions, you sometimes see a band of warmer air above a band of colder air. This is known as an inversion, and the warmer air will reflect signals in the VHF/UHF range back to Earth. As a matter of fact, the higher the frequency, the better this works! So, Tropospheric Ducting is a big boom for 70-centimeter operators. Signals can travel hundreds of miles during these inversions!

Normally, weather such as warm days with cool nights indicates conditions favorable for Tropospheric Ducting. Also, hazy days are a reasonably good indicator, as warm air tends to trap haze and/or smog, often revealing a satisfactory setting for inversions.

Ground Wave

A form of propagation that you will hear referred to from time to time is Ground Wave. This is basically the ability of the lower frequencies to hug the Earth's surface as they stretch out away from the transmitting station. Normally, this propagation is limited to a few hundred miles, but as with everything else, it can be greater when conditions permit. Ground wave is seen, however, almost exclusively in the HF spectrum. The VHF/UHF regions experience virtually none of this propagation due to their frequencies.

On the HF bands, ground wave can be very useful for making short-distance contacts. Normally you would think that the closer a receiving station is to the transmitter, the stronger the received signal would be. This sounds logical, but in reality, close stations are often difficult, if not impossible, to contact. What happens is our signals normally project toward the ionosphere, bounce off it, and back to the ground. This is what is referred to as skip, and the area between the transmitter and the receiver is the skip zone. Unfortunately, whatever is below that zone receives very little, if any, signal from the transmitter.

Hence, those close stations don't even hear our signals, UNLESS ground wave propagation is in effect. In that scenario, the hugging dynamics of ground wave keep the signal close to the ground and it can then be received by nearby stations.

Next, let's look at some other natural occurrences that lead to admirable propagation above 30 megahertz. Probably the two most important phenomena are Auroras and meteor showers. Both of these will make working the VHF and UHF bands a much more pleasurable experience.

Auroras

Auroras are the product of solar flares that saturate the atmosphere with ultraviolet energy. This energy tends to collect at Earth's poles and is responsible for the vivid displays of light and color known as the Northern Lights. Naturally, if you are far enough north to observe these lights, this is an excellent indicator that Aurora propagation might be in effect. One interesting aspect of this type of propagation is that you will need to bounce your signal off the polar regions to achieve the effect. But, for some long-distance 2-meter communication, albeit probably fairly short in duration, the Auroras can be great!

Meteor Showers (Meteor Scatter)

Meteor showers help us out in a slightly different way. As meteors enter the Earth's atmosphere, they burn up. That is a result of the friction generated as they encounter the gases in our atmosphere. This process leaves a trail of ionized air and charged particles, and if the trail is dense enough, VHF signals will reflect off it. 6 meters benefits the most from meteor showers, but it can also aid 2-meter communications.

Hence, meteor showers — sometimes referred to as meteor scatter — can be of very real assistance to long-distance propagation in the VHF bands. Since it is hard to predict exactly when these showers are going to occur, it is best to monitor your local news or check the Internet for times they are scheduled to happen. In that fashion, you may well be able to take advantage of this type of propagation.

Solar Flares

Solar flares are the next subject to consider. Strong energy levels from the sun tend to make long-distance communication better, but the sun can overdo the program. And, this is one area where it tends to do just that. Flares from the sun can really disrupt the Earth's magnetic fields, so a big event regarding solar flares is not a good thing. One of the ways to track the sun's activity is to check the various amateur radio-related Web sites on the Internet. Naturally, you do have to have a computer and Internet access, or a friend who does. But, these will provide a good deal of information concerning solar action. It will also provide other highly pragmatic data about propagation in general. One excellent source of this type of information is the National Aeronautic and Space Administration (NASA) site at www.nasa.gov. Between these two references, you should be able to find what you need!

Evaluating the Data

Among the intelligence you will encounter are the venerable K and A indices, both measurements of geomagnetic activity. The K index is taken every three hours, eight times daily, and the average of the eight daily measurements becomes the A index. What does this mean to you? It will tell you how much geomagnetic activity is occurring, and that is a blueprint for the type of propagation you can expect. Normally, the lower these indices, the better, with both at zero being perfect. But, as they say, nothing is perfect!

Sidebar here! I have been writing books for five or six years now, and I try to ask this question in every text I've written. Who is "THEY?" I have yet to get an answer, so if you know, please pass it on to me!

Sorry about that! It is a little pet peeve of mine (you know, a small fury thing with long sharp teeth). Anyway, where was I? Oh yeah, this business of nothing being perfect applies to propagation as well. What you might see as a report on one of the amateur radio sites could look like this: "WWV SFI = 178, A = 33, K = 4, Low/Mod; Geo − Unset/ Minor Storm Level." To translate this, it is telling you that the Solar Flux Index (SFI, or ionization level) is at 178, the A index is 33, and the K index is 4. "Low/Mod" refers to the geomagnetic activity that is rated at "low to moderate." Further, the geomagnetic activity is "unsettled" (Unset) with minor geomagnetic storm levels.

An honest evaluation of this particular point in time would be modest DX success. The bands would be open, but not at their highest level of activity. Lower A and K indices and a higher SFI would definitely improve band conditions (although a solar flux of 178 is not bad). However, unsettled geomagnetic activity does not help matters. In this report, there is no mention of the sunspot numbers, and you want that to be a high number! But, since we do not have that data, it can't be considered in our evaluation.

What to Expect

So, you now have a firm comprehension of propagation. You do, don't you? Of course you do! I mean, what else? Okay, let's talk a little about what to expect on the various bands assigned to the amateur radio service.

High Frequency (HF)

Concerning the high-frequency (HF) bands, let's start at the bottom with the 160- (1800 to 2000 kilohertz) and the 80-meter (3500 to 4000 kilohertz) bands. These are both nighttime bands, where sky wave propagation can carry signals thousands of miles during the winter months and several hundred miles in the summer. Daytime distances are limited to around 150 miles regardless of the time of year.

The 80-meter band will show a little better propagation than 160 meters, but not much. However, contrary to what has been said about sunspots, both bands are better during periods of lower sunspot numbers than higher numbers. This is due to the fact that ionization resulting from the sunspots tends to absorb signals at these frequencies. Hence, they become valuable when the solar cycle is in the down segment.

Moving up to the 40- (7000 to 7300 kilohertz) and 30- (10.1 to 10.15 megahertz) meter bands, these two are also similar in behavior. Both provide better signal distances during daylight than at night, with the winter months furnishing the best propagation. Also, 30 meters shows better performance over 40 meters than 80 does over 160. One major problem with 40 meters, especially at night, is foreign broadcast stations. These stations transmit with high power levels and can wipe out 40 at times. This is notably true on the upper end of the band. However, when this interference (QRM) is not present, 40 meters can be a great band for U.S. and Canadian traffic.

The 30-meter band is restricted by the FCC to CW and digital only. However, it doesn't suffer from the interference found on 40 meters and is dependable for both daylight and nighttime communications. Again, both these bands endure some degradation from high sunspot activity, but nothing compared to 80 and 160 meters.

Next, I'm going to lump 20 (14.0 to 14.35 megahertz), 17 (18.068 to 18.11 megahertz), and 15 (21.0 to 21.45 megahertz) meters together. There are some differences in these bands, but for the most part they behave very similarly. All three are great DX bands and probably display the most open time of any of the amateur radio HF frequencies.

20 meters is by far the most popular of these bands. For that matter, it is also one of the most popular HF bands. This is due largely to the fact that 20 is often at least partially open just about anytime. This is especially true during peak solar conditions. Thus, you will be able to work it year-round.

17 meters is one of the WARC bands. WARC stands for World Authority Radio Conference, an organization that allocates frequency spectrum. Awhile back, WARC authorized worldwide amateur radio activity on these frequencies. On a final note, 17 acts much like 20 meters, with fewer daylight openings during low sunspot activity periods.

The 15-meter band is again similar to 20 meters, except that it suffers far more from low sunspot numbers. You may have noticed that as these frequencies increase, their dependence on good solar activity also increases. This is just the nature of radio frequency energy. During the down portion of the cycle, 15 meters tends to close out. However, it can be even better than 20 meters when the cycle is peaking.

To round out the HF region, let's look at 12 (24.89 to 24.99 megahertz) and 10 (28.0 to 29.7 megahertz) meters. Like the other band pairs and the triad, these two bands exhibit very similar properties. Both are highly reliant on good sunspot activity and are virtually dead during low cycle conditions. During peak cycle periods, 12 meters, also a WARC band, can be used for worldwide DX , and 10 meters is nothing short of breath-taking. 10 is home to a number of DX and HAM-related networks and is a favorite among many operators when conditions are favorable. And don't forget, 10 meters takes advantage of Sporadic-E from time to time!

VHF and UHF

Moving up the spectrum ladder, we come to the 6-meter band (50 to 54 megahertz). This is often called the magic band, but is just as often referred to by other names I can't repeat in this text. After all, this is a family book produced by a family publisher! In all seriousness, 6 meters is a strange place! And, I don't mean that in a negative sense! I simply mean that it is hard to predict what this band is going to do next. For example, during peak cycle conditions, communications for hundreds, even thousands of miles is possible. But, when the cycle is down, it is good for little more than local communications. That is what you would expect, right? However, if Sporadic-E or Tropospheric Ducting is in effect, then 6 meters is back to a reasonably long-distance status.

This band is also the highest frequency that propagation by means of reflection from the F layer is observed. In that lies some of the secrets behind 6 meter's ability to travel long distances. So, in some ways, 6 meters enjoys the best of several worlds. Additionally, it is a popular band for local repeater operation, as the direct wave nature of the signals is conducive to local communications just about anytime.

Michael KG4KSG works a 2-meter repeater with his mobile rig.

Next comes the revered 2-meter (144 to 148 megahertz) band! 2 meters is unquestionably the most popular amateur radio band in the world. I say that without pause or reservation! You will be hard pressed to find a more active HAM band anywhere you travel, but there are some good reasons for this.

First, being a non-HF band in most countries, 2 meters is a privilege that comes with an entry-level amateur radio license. Second, the properties of these frequencies are superbly suited to short-range or local communications (ragchewin'). And last, but certainly not least, its popularity has led to mass production of radios, which has led to cheaper equipment prices! So, for a reasonable expense, you can set up a mobile and/or base radio station and be on the air. Also, radio clubs make good use of 2 meters for repeaters that help extend transmission range, especially of the handheld transceivers (HTs). But, even with repeaters, the range is rarely over 100 miles and that is with a repeater installed high on a commercial tower (1,500 feet or more).

As for propagation, 2 meters is pretty much line of sight. That is, except during times of abnormal atmospheric conditions, it isn't going to travel much over the horizon. It is very reliable in this capacity, though, as there is substantially less noise at these frequencies than in the HF region. And, this is another reason for the popularity of the 2-meter band.

About those abnormal conditions! We have already discussed this, but let me refresh your memory a little. During periods of Sporadic-E, Troposheric Ducting, meteor showers, or Auroras, the chance of 2-meter signals traveling longer distances is greatly increased. So, keep an eye out for such conditions.

This next band, 1.25 meters (222 to 225 megahertz) is sadly neglected in many parts of the United States. Here is a band that enjoys much of the tranquility of UHF in terms of noise and yet maintains a lot of the carry power of 2 meters. Despite this, it is hardly used in many areas. In fact, at one time we had 220 to 225 megahertz, but there was so little activity on this band that we lost 220 to 222 megahertz.

Part of this could be a lack of inexpensive gear, as manufacturers simply don't sell enough of these radios to allow the prices to be competitive with 2-meter rigs. Unfortunately, if the band is not being used in a certain area, there is little incentive for HAMs to buy equipment capable of operating in that range. So, the future of 1.25 meters is questionable at best. I hope we don't lose it altogether, but with the competition these days for VHF spectrum being what it is, I fear commercial concerns will eventually gain control of this band.

Propagation on 1.25 meters is similar to 2 meters — pretty much local communications. This band is also used for repeaters and often chosen for "links" (dedicated connections) to other repeaters or receiving sites. During Sporadic-E events, 1.25 meters can benefit, but usually not as much as 2 meters. Tropospheric Ducting, on the other hand, will bring much better propagation to 1.25 meters than it will to 2 meters.

Another popular local transmission band is 70 centimeters (420 to 450 megahertz). Often referred to as "440" (due to the initial frequency of the phone portion of the band), on the average, 70 centimeters does not experience as much activity as 2 meters. But, in many places, such as large metropolitan areas where repeater spectrum is at a minimum, this band sees heavy activity. Additionally, these UHF frequencies are great for handheld equipment when only short-distance communications is needed.

70 centimeters is also home to some of the more exotic aspects of HAM radio. Such things as "moon bouncing," amateur television, and satellite contacts all make good use of this band. Being purely line of sight, normally direct communication distances will be shorter than with 2 meters, but the quiet nature of 70 centimeters offers a

strong inducement to using this band. As far as propagation goes, Tropospheric Ducting is about the only advantage this band will experience.

The last UHF bands I'm going to discuss are 33-centimeter (902 to 928 megahertz) and 23-centimeter (1240 to 1300 megahertz). Be reminded, however, that we as HAMs have privileges on bands way up in the gigahertz region, but most of us rarely, if ever, use those frequencies. The properties of the 23- and 33-centimeter bands are very similar. They are on the edge and in the microwave region of the radio spectrum, respectively. They are also used primarily for very short line-of-sight communications and/or experimental purposes. You are not going to find too many HAMs on these bands that are not hard-core enthusiasts. Here, even local contacts can be questionable due to the high reflectivity of these frequencies. These signals will bounce off just about anything!

Another problem with these bands is the threat of personal injury. As you may know, radio frequency energy will cause heating of organic tissue, and this is true, to some extent, at any frequency. But, at these wavelengths, this becomes an increasingly serious problem. Remember, both bands are, for all intents and purposes, in the microwave range, and I doubt I need to remind you what a microwave oven will do to organic tissue!

Propagation is seen only in the form of Tropospheric Ducting, and even then, it's not all that great. Some very long-distance contacts have been made (200 to 300 miles), but these were accomplished by two HAMs pointing dish-style antennas at each other from the top of a couple of mountains. Basically, these frequencies are only useful for short-range communications and/or data transfer and experimental gear.

Conclusion

Well, there you have a brief, but hopefully somewhat sufficient, explanation of propagation. A lot is not covered here, and you may well want to look deeper into this subject. But, what I have included are the primary factors that influence the way radio waves behave in our atmosphere. The Internet, and library are both excellent sources for additional, and more in-depth information regarding this subject.

At first meeting, propagation can be downright intimidating! However, as you delve deeper into it, it becomes both an interesting topic, and a great aid in helping you get better performance out of your radio gear. Additionally, as you work the bands, much of what has been said here will suddenly sound familiar to you and your understanding of this sometimes baffling aspect of HAM radio will improve. In short, don't be afraid of propagation, as it is your friend!

~ Section 4 ~

What Do You Say?

Let's Build Some Stuff!

Projects

Introduction

Personally, I love to build electronics projects, especially from scratch. I found though, that not all HAMs share my enthusiasm for this pastime. I must admit that surprised me a little when I first got into HAM radio. I had some background in the hobby from past friends, and they all were avid project builders. Amateur radio has changed a lot since those days, and apparently so has the homebrew fever that once was so prevalent.

However, I decided it was still a good idea to include some simple projects in the text. Maybe, these will encourage HAMs who aren't project inclined to give it a try. Building your own equipment is a lot of fun, and a certain pride is associated with making a contact on a 40-meter QRP station you constructed with your own two hands (and a soldering iron and some tools, of course).

To that end, I have selected 10 goodies that I think you will like. Each one takes a little skill in homebrew, some more than others, but all are easily within the capability of anyone already involved or interested in this hobby. Really, I have no doubt that anyone reading this book can build this stuff.

As I always like to do in my books, I have included one project that is a kit. Kits were big in my early electronic days, so I guess this is kind of a nostalgia thing, at least for me! Kits are a very handy way to obtain all the parts and instructions you needed to build that special piece of equipment.

Without further delay, let's get started. Try some of these out as I believe you will really like them, maybe even more than you thought you would!

Project #1

Handy Dandy Field Strength Meter (FSM)

Introduction

This little gem is always nice to have around. It doesn't require any connection to your radios, but reads the relative strength of their signals. Not only does it let you know your transmitter is working, it also tells you how well it is working.

Once calibrated, this field strength meter (FSM) can be used to take measurements for your routine station evaluations and keep you legal with the FCC. It also proves its worth when testing equipment and/or tracking down signals of unknown origin or location. With that said, let's take a closer look at this relatively simple circuit. With an initial investment of $20 in components (the meter movement is the most expensive part), a few hours of work and you have an addition to your shack that is quickly appreciated.

Theory

What we have here is a device that detects a radio frequency (RF) signal, amplifies it, then sends it to a meter movement where its strength is displayed. This type of circuit is also known as a passive detector, because unlike many receivers, it does not emit any form of signal itself.

Most receiving equipment employs front ends (detectors) that are oscillators and radiate a signal. This is why most airlines do not allow you to listen to a portable radio when in flight. That oscillator signal might interfere with the aircraft's on-board communications and avionics gear.

Looking at the schematic (Figure 1-1), you see the antenna is connected to the junction of a pair of germanium diodes wired in series. These diodes (D1 and D2) and capacitor C1 form the rudimentary detector circuit for our FSM. By the way, these have to be germanium diodes, as the more common silicon variety does not work as a detector.

The C1/D1 junction is then fed to one leg of a potentiometer R1. R1 is used to calibrate the circuit by controlling the amount of signal that reaches the transistor amplifier. That amplifier consists of transistor Q1 and resistors R2, R3 and R4. These components are configured as a standard amplifier with the resistors regulating the bias voltages.

The signal from the potentiometer's center lead, or whipper, is applied to Q1's base. In this fashion, that signal controls the collector/emitter flow within the transistor. The higher the detector signal, the more flow, hence a higher reading on the meter (M1).

Power for the circuit can come from a battery if you want portability or from your base station's power supply. I have mine connected to the same supply that runs the transceivers. Also, check the surplus market for lighted meters. These movements include small grain-of-wheat incandescent bulbs that light up the meter, which is a nice feature.

That is about all there is to this device. When you transmit, the antenna picks up enough of your signal to activate the detector, which in turn drives the amplifier and meter and the meter's needle advances. How far up the scale it advances depends on the potentiometer setting and transmitter power. That is explained in detail later.

Figure 1-1: Field Strength Meter Schematic

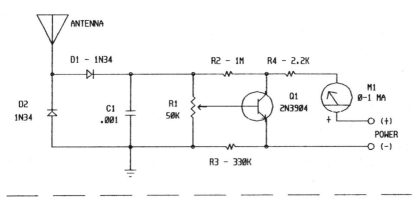

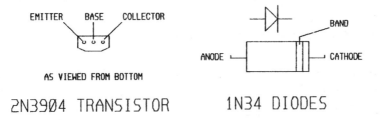

Construction

This project can be constructed with either point-to-point wiring, on perforated construction stock or, if you have the capability, make your own printed circuit board (PCB). If you have read any of my other project books, then you know I'm a nut for PCB construction. I find it easier to work with and far more reliable. However, not everyone can make his or her own boards, so the perf-board method works also.

The first step is to install all the discrete components such as capacitors and resistors on the board (see Table 3-1). None of these parts are polarity sensitive. Next, install the diodes and transistor. Here, you do have polarity, so be sure to observe that factor on the diodes and the orientation on the transistor (see the schematic).

I always like to put the semiconductors on last, as they are somewhat heat sensitive. While most transistors, diodes, and integrated circuits (IC) are designed to handle the normal amount of heat necessary for installation, it's not a good idea to expose them to anymore heat than required. When they go on last they are not getting heat from the installation of other components.

With everything on the board, solder appropriate lengths of hook-up wire for the meter movement as well as antenna and power connections. When I say appropriate lengths, I mean enough wire to reach the various components or connections, but not an excess. For the prototype, I placed the circuit board and meter in one case, so only power has to be fed in with hook-up wires. The antenna is just a short (30-inch) length of wire. The power leads are kept to a length that easily reaches the power supply, but not so long as to get in the way.

Project 1: The completed Field Strength Meter (FSM) featuring a lighted meter.

Testing and Calibration

Once that is done, the unit is ready to test and calibrate. First apply power, then using an HT or other transmitter, send a test signal. If all is well, you can see some deflection of the meter's needle. As a matter of fact, you may see too much deflection and the needle may pin (deflect fully). Be careful of this, as those needles are very delicate and easily bent. Full deflection can slam them against the stop posts and damage the movement. Next, adjust potentiometer R1 back and forth while the test signal is present. Substantial change in the needle position is observed as you do this.

To calibrate your FSM requires another already calibrated meter. You should be able to borrow one from a fellow HAM. With both meters in hand, again send a test signal and notice the reading you get on the calibrated FSM.

Now, it is just a matter of setting the potentiometer in your meter to obtain the same reading with that same test signal. Once the setting is correct, you shouldn't have to touch it again. Only a change in the power supply, or antenna length requires recalibration.

Now you have a field strength meter ready to use for various signal measuring duties. You built it yourself. One nice aspect concerning this last part is you can fashion the unit in any configuration you want. With factory-built equipment, you are stuck with that manufacturer's design. That is not to say that much of the commercially built equipment isn't nice in style and appearance. Most companies put a lot of effort into the image they project to the public. However, it is great to be able to design and build a piece of gear that totally meets your needs and standards.

Conclusion

There you have the Field Strength Meter. This is a one night project for even an inexperienced builder. It comes in mighty handy. Trust me!

As has been stated before, HAMs operating transmitters of 50 watts or more output have to keep track of possible radiation hazards emanating from their stations. With this nifty little device in your arsenal, that isn't a problem. In fact, you can check it as often as you want.

So, as a first project for your new station, if you are setting up a new station, or as an addition to an established shack, this FSM is ideal. Not only does it provide a highly pragmatic piece of equipment, but also an enjoyable evening of hands-on work with a little learning about electronics thrown in for good measure.

Have Fun!

Table 1-1

Field Strength Meter Parts List

Semiconductors

Q1-2N3904 NPN Signal Transistor
D1-2-1N34 Germanium Diodes

Resistors

R1-50,000 Ohm PCB Potentiometer
R2-1,000,000 Ohm 1/4 Watt Resistor
R3-330,000 Ohm 1/4 Watt Resistor
R4-2,200 Ohm 1/4 Watt Resistor

Capacitor

C1-0.001 Microfarad Disk Capacitor

Other Components

ANT-30 to 40 Inches of Wire (This will depend on the strength of the RF in and around your station).
M1-0 to 1 Milliamp-Full Scale Analog Meter

Miscellaneous

Solder
Hook-up Wire
Case (if desired)
Hardware
Circuit Board Materials
Power Connection Materials

Project #2

A Radio Frequency (RF) Relay

Introduction

This is a handy device much along the same lines as the previous project. It utilizes a passive detector to sense the presence of radio frequency energy. Once the circuit perceives that energy, it activates a relay that, in turn, can perform a number of useful tasks.

One task I have for my radio frequency relay is to mute the scanner when I transmit. That way I don't get that annoying squeal associated with audio feedback. Also, the relay activates an "on the air" sign.

If you are using a scanner as part of your shack, this unit proves its worth rather quickly. It can also be used to turn on/off any other device when a strong (close by) RF signal is present.

Theory

Looking at the schematic (Figure 2-1), this circuit is different from the field strength meter, but not by much. Mainly, it is less complex, which should be good news. The primary changes are a DC blocking capacitor on the antenna and using the transistor as a switch instead of an amplifier.

Concerning the first distinction, in this application you want to block DC from getting into the circuit, which could result in false triggering. Capacitor C1 in series with the antenna does this job nicely. That is not a problem with the FSM as it only reacts to radio waves (AC).

From C1, diodes D1 and D2 handle detection duties. Again, these have to be germanium diodes, as the more common silicon rectifiers are unable to detect RF. The detected signal is then fed to one leg of potentiometer R1. This, as in the last circuit, provides sensitivity by restricting the amount of signal that reaches the transistor.

Transistor Q1 acts as a switch to activate the relay (turn it on and off). When a potential is applied to the base of Q1, electricity flows from the power source, through the collector/emitter junction and to the coil of relay RLY1. That generates a magnetic field that closes/opens (depending on how you have them wired) the relay contacts. Capacitor C2 is used merely to filter out the ripple and make the circuit more efficient.

The power source can be either a battery for portable use, or a standard AC power supply. Voltage is not critical, as the transistor handles up to about 40 volts. However, you want to use a relay that is rated for whatever voltage you employ. For example, if you decide to use the 13.8-volt supply that powers your equipment, then a 12-volt relay would be in order.

A little overvoltage does not hurt anything, in fact it may result in a more positive relay action. As for the contacts, relays come with a multitude of different configurations. You can get everything from SPST, SPDT, DPDT to 4PDT, 8PDT, or DP8T and so on. Hence, a number of devices can be controlled by a single relay.

Figure 2-1: Radio Frequency (RF) Relay

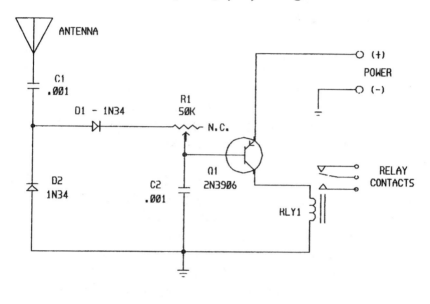

ORIENTATION OF THE SEMICONDUCTORS

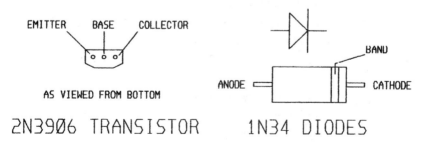

This is a simple yet highly pragmatic circuit just waiting for the chance to serve you. Now, let's move on to putting this gem together.

Construction

As with the field strength meter, the RF relay can be built with either printed circuit board (PCB) or point-to-point wiring on perf-board. I like PCB, but it really is a matter of personal preference. After all, you have just as much right to my opinion as I have! Think about it!

With your board in hand, install the discrete components (C1, C2 and R1) first, then the relay RLY1. Next, add the diodes D1 and D2 and transistor Q1. Be careful to observe polarity and orientation on these semiconductors! Last, solder power hook-up wires and an antenna to the board. Power lead lengths should be whatever is necessary, while antenna length depends on how sensitive you want the relay. The longer the antenna, the more sensitivity the circuit has. A procedure for determining antenna length is covered shortly.

That, as they say, is that! Very simple circuit to construct. If you plan on putting the relay in an enclosure, you could be thinking about that now. I wouldn't seal it up until you have tested it, and that is coming up next.

Test and Operation

Testing this device is about as simple as making it. Apply power to the circuit and see if the relay clicks. If it doesn't, leave potentiometer R1 as it is. If RLY1 activates, adjust R1 until the relay just releases. Now, key a transmitter and again check for relay activation.

If nothing happens, either R1 is set too low and needs re-adjusting, or the antenna is not long enough. First try changing the R1 setting while the RF signal is present. Be sure the potentiometer is run through its entire range. That may do the trick. If not, then the antenna is the problem.

To determine how long of an antenna is needed, place the relay in its permanent location, then temporarily hook a length of wire to the antenna connection. Start with about three feet. If the relay functions properly when a transmitter is keyed, you might want to leave well enough alone.

However, if that length of antenna is inconvenient, begin trimming it off in half-inch segments. Eventually, you reach a length where the relay no longer activates. At that point, you want to add an inch, cut a new piece of wire to that length and install that wire as the permanent antenna.

If the relay does not activate with the three foot antenna, go the other way, and add lengths of wire until you get relay action. Here, it is best to increment in one foot lengths, then once a functional antenna is obtained, trim it down as before.

If you are still having trouble, be sure to check the orientation of the semiconductors, and as silly as it sounds, that you do have power to the circuit. You would be amazed how often the power supply is the culprit. I remember once talking to a TV repairman about this problem. He said it wasn't unusual to arrive at someone's house (that was back when they actually came to your house) on a repair call only to find the set unplugged.

The moral to that story is, always check the power source when encountering difficulty. It may be putting out juice, but not enough, it may be dead altogether or, it may be unplugged.

Now that we have our relay tested (we do, don't we?), all that is left to do is hook up whatever device(s) we want to control with our transmitter's RF signal. As I stated earlier, I have used my relay to cut off the audio on a scanner that is monitoring HAM frequencies to prevent feedback.

To do this, your relay needs DPDT relay contacts, as you want to direct the scanner's audio output from the speaker to an 8-ohm resistor. The purpose here is to apply an 8-ohm load to that output, thus preventing any damage. It is never a good idea to run power amplifiers without a load. Doing so can burn up the final output stage.

Project 2: The completed Radio Frequency (RF) Relay circuit board. This is a perf-board layout without the leads connected.

Again, as stated, I have also used the relay to light a sign that reads "on the air." This is a handy and conspicuous way to let others know you have your transmitter keyed. However, this is only one auxiliary device that could be activated/deactivated by the relay. If you think about it, I'll bet you can come up with at least one thing you would like to control.

Conclusion

If you want to cut off your scanner, light up your life and HAM shack, or just manage some other piece of equipment with your transmissions, this is your boy! It also illustrates an electronic principle that might come in handy later.

Sampling, or sniffing RF signals does have its merit. In addition to indicating the presence of radio frequency energy in and around your shack, it can also find such energy elsewhere. That makes the RF relay handy for locating signals of unknown origin and/or position.

Try this one out. I'm certain you will find it useful and instructive. If nothing else, it will make you look very intelligent when people ask how you did that, and you can tell them.

Table 2-1

Radio Frequency Relay Parts List

Semiconductors

Q1-2N3906 PNP Signal Transistor
D1-2-1N34 Germanium Diodes

Resistors

R1-50,000-Ohm PCB Potentiometer

Capacitors

C1-2-0.001 Microfarad Disk Capacitors

Other Components

ANT- 30 to 40 Inches of Wire (this depends on the strength
of the RF in and around your station)
RLY1-DC Relay (the coil voltage depends on the voltage of
the power supply used)

Miscellaneous

Solder
Hook-Up Wire
Circuit Board Materials
Case (if desired)
Hardware
Power Connections Materials

Project #3

Power Distribution Center

Introduction

This project comes to us courtesy of Mike KT4XL and is a dandy addition to any HAM shack. Mike calls this a power distribution center (PDC), and it allows you to connect multiple radios (equipment) to a single power supply.

With most amateur radio power supplies, the binding posts that deliver the output are designed for one wire each. Now, on the surface, that might not sound like that big of a deal. However, once you experience the joy of trying to connect more than one wire to these posts, I can almost guarantee you will develop a deep appreciation for Mike's PDC.

With the power distribution center, the only binding post connections for the power supply are the lines to the PDC. Once the PDC is installed, you have a whole row of separate posts to which you may connect various pieces of equipment.

While that in itself is outstanding, there is more! Too often one or more of those multiple leads that you have squeezed under the posts works its way loose. This not only makes for a bad connection, but worse, can short your power supply. Trust me on this, you don't want that to happen. Such a condition can put a tragic, if not spectacular, end to your supply in an enormous hurry.

So, since I know you don't want that to happen to your power supply, I strongly suggest you build a power distribution center. Not only does the PDC protect you against such a disaster, it also makes your life easier when it comes to connecting equipment.

Theory

The PDC can be configured in a variety of ways, but all layouts share one thing in common. Everything is wired in parallel. Looking at the schematic (Figure 3-1), you can see what I mean. In the prototype for this section, Mike included a light emitting diode (LED) power-on indicator, a 0 to 15-volt DC analog meter and a small grain-of-wheat incandescent bulb.

The meter informs you of the voltage present on the distribution center, and the small bulb is used to illuminate that meter in subdued lighting conditions. The only series component is a 40-amp fuse in-line with the positive lead of the PDC. That, of course, is protection against overload.

Hence, with the various components wired in parallel and contained in a suitable enclosure (an elongated box), the power distribution center is ready for service. You can add as many sets of binding posts as you feel are needed (Tip: over-estimate this, as new equipment is always a BIG option with HAMs). All in all, a very simple arrangement for such a useful device. In fact, they don't get much more simple than this. That makes construction easy.

Figure 3-1: Power Distribution Center

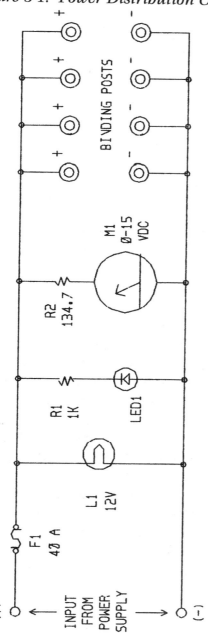

THIS DIAGRAM SHOWS FOUR PAIRS OF BINDING POSTS, BUT YOU CAN ADD AS MANY PAIRS AS YOU WANT OR NEED. THIS CONFIGURATION IS DESIGNED FOR A POWER SUPPLY THAT PROVIDES 12 TO 14 VOLTS DC. AGAIN, CHANGES IN THE COMPONENT RATINGS WILL ALLOW DIFFERENT POWER SUPPLY LEVELS.

Construction

As stated, you need an elongated plastic box (metal can be used, but be VERY careful of shorts) large enough to house your planned configuration. While the meter, LED, and small bulb are nice to have, they are not essential and can be left out if desired.

Presuming, however, you intend to build your PDC as Mike did, start by drilling and/or cutting the necessary holes for the binding posts, LED and meter in the front panel. You also need a hole, in the back or on one side of the box, large enough for the 10-gauge power supply connection leads.

Once that is done, install and wire the various panel components. To connect the binding posts, Mike used pieces of double-sided PCB material as rails (one on each side). This is a good idea, as it provides a heavy-duty bus line

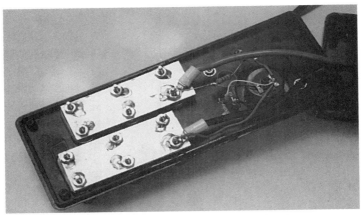

Project 3: Internal view of the Power Distribution Center (PDC). Note the PCB material strips on the left used to connect the binding posts.

and also acts to dissipate any heat that might be generated at the posts.

If you don't go that route, use 10-gauge stranded hook-up
wire between each post. Naturally, all the positive and all
the negative posts are tied together.

Next, you want to connect the LED/R2 assembly to the
appropriate lines. Hold off on the meter until we get the
bulb installed. To do that, remove the clear plastic front
from the meter. This gives you access to the face plate. Be
careful with the meter's pointer as it is very delicate. Drill a
small hole in the faceplate in one of the bottom corners
(that way, the bulb is hidden when the plastic front is put
back in place) just large enough for the bulb. Secure the
bulb with cement. Don't use epoxy for this, as the bulb
gets hot, and epoxy tends to melt.

Once that is completed, the plastic front can be replaced,
and the grain-of-wheat bulb wired to the meter connection
points. Since the bulb has no polarity, it doesn't matter
which lead goes to which connection post. Incidentally, resistor R1 comes with the meter if you use the one mentioned in the parts list.

Now, the meter can be placed on the front panel and wired
to the circuit. Make sure you observe the meter's polarity,
as it reads backward if the connections are reversed.

The last step is to hook up the two lines that go to the
power supply. Again, they are connected to the positive
and negative rails, and it is best to use color-coded wire,
red for positive and black for negative. Also, this cable needs
to be 10-gauge or larger to handle the amperage from the
power supply.

For the fuse, Mike used a 40-amp automobile spade fuse set in an in-line holder. It is hard to find standard fuses that large, but if you do find one, that approach also works. The fuse is wired in series with the positive power supply lead.

That is about all there is to it. Once you have double-checked your work, close up the case, and the PDC is ready. It would be a good idea to place an ohm meter across the two input leads just to make sure there isn't a direct short. You wouldn't want to subject your power supply to that!

Test and Use

The only real test for this project is to hook it to your power supply, and see if you get the expected voltage on the meter (if you included one) and at each pair of binding posts. Also, the LED and meter bulb should both light. If that checks out, all is well.

For use, that is simple. With the PDC connected to the power supply, you merely hook up each piece of equipment to one of the binding post sets. In this fashion, you have power on everything connected to the power distribution center.

You could incorporate a power switch in series with the fuse/positive lead, but that might not be such a good idea. It is best to turn off each radio before killing the main power source. Doing so protects the equipment from surges when the power is initially restored. However, if you're like me and that all-in-one switch is available, you will probably get lazy and opt to use that instead.

Conclusion

This is a one evening project that is extremely handy, especially as you expand your shack. In addition to being simple to build, it can also be quite inexpensive — you know, CHEAP! BOY, that always gets my attention.

It can, however, be more elaborate if you wish. You could include both a volt meter and an amp meter to keep tabs on each of those quantities. This increases the cost, but it might well be worth it if a knowledge of those values is important.

No matter how you configure the power distribution center, I'm certain you will enjoy using it. The PDC is one of those additions to your shack that proves valuable time and time again.

Table 3-1

Power Distribution Center Parts List

Semiconductors

LED1-Red or Green Light Emitting Diode

Resistors

R1-1,000-Ohm 1/4 Watt Resistor

R2-137.7-Ohm Precision Resistor (this will come with the meter if you use the meter listed below)

Other Components

L1-12 Volt Grain-of-Wheat Incandescent Bulb
M1-0 to 15-Volt DC Analog Meter (Radio Shack: 22-410)

Miscellaneous

Solder
Hook-Up Wire
Heavy-Duty (High Amperage) Binding Posts
Elongated Case
Power Connection Wire (10-gauge or larger)
Hardware (LED Holder, etc.)

Project #4

A Simple Ground Plane Antenna

Introduction

I don't know who came up with this admirable method of building a ground plane antenna, but I would like to express my appreciation to that person. You know who you are! Thank you! Thank you! Thank you!

Really, this technique is ideal for putting together a unity gain (no gain) ground plane-style antenna for 2 meters and 70 centimeters, and, for that matter, any of the higher frequency bands. The parts count on this one is nothing short of ridiculous: a PL-239 female UHF connector, four bolts and five lengths of 1/8-inch steel rod. You can even use 8- to 12-gauge solid wire instead of the rods, if that catches your fancy.

The end result is a highly efficient antenna. It doesn't have any gain, but for local 2-meter ragchewing or scanner reception, it is hard to beat. At least, considering its cost!

So, let's dive into this project with vigor! It might consume two hours of your time, but that's if you take a few naps along the way or work very slowly. I know you won't want to do that, though, as you will surely wish to get this gem finished as soon as possible. Right? Right!

Construction

For this project, we are going to build a 1/4-wave antenna for the 2-meter band (144 to 148 megahertz). The elements will be 1/4 the length of a full wave for the band. Our first task is to cut the steel rods to their correct length. That length is determined by the formula, and this one is simple: 234/the frequency in megahertz, or 234/146. I used 146 as the frequency because it is the center point of the 2-meter band.

That gives us an element length of 19¼ inches. However, every description of this antenna I have seen, recommends 19 inches for the center radiator and 19¼ inches for the four radials. I don't know why, but it works and that is how I designed this prototype.

Project 4: The necessary components to build the 2-meter Ground Plane Antenna: five metal rods and an SO-239 antenna connector.

With the elements cut, the four 19¼-inch rods are secured with the bolts to the four mounting holes of the SO-239 connector. Probably the hardest part of this process, and for that matter the entire project, is bending the ends of the rods to conform to the bolts. In fact, some of the articles recommended soldering the rods to the 239. However, that makes for a less sturdy antenna, so I opted for the bolts.

Once the radials are in position, starting at the connector, bend them downward to an angle of approximately 45 degrees. This helps adjust the antenna impedance to the required 50 ohms.

Now, it is time to solder the remaining 19 inch rod to the SO-239's center solder lug. Use enough solder to make this secure, but not an excess. Between this solder joint and the bolts, the antenna is rugged enough for outdoor use.

As for using the antenna, the feed line is comprised of a length of 50-ohm coaxial cable with a PL-259 male connector on one end and whatever connector the radio needs (BNC, N, etc.) on the other end. Connect the 259 to the antenna's SO-239 and the opposite end to your radio (see, that's simple enough).

To reduce line loss, try to keep the feed line as short as possible. Runs (lengths of coax) shorter than 50 feet can probably use the lighter RG-58 coax.

Well guys and gals, that's it! A bracket can be fashioned that slips over the threaded part of the SO-239 for mounting the antenna to a mast or once the coaxial connector (PL-259) is in place, cable ties can be used to strap the unit to whatever it is you want to strap it to.

The diagrams and photographs show more detail regarding this assembly, but this is about as simple as it gets — when it comes to antennas, that is. For other bands (frequencies), simply employ the 234/frequency in megahertz equation, and you to get the lengths you need for the elements.

Figure 4-1: The 2-Meter Ground Plane Antenna

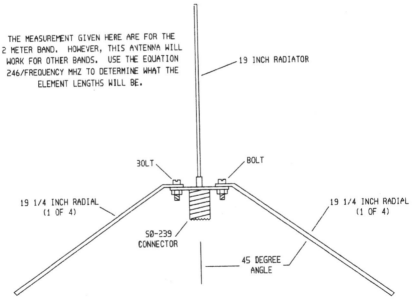

THE MEASUREMENT GIVEN HERE ARE FOR THE
2 METER BAND. HOWEVER, THIS ANTENNA WILL
WORK FOR OTHER BANDS. USE THE EQUATION
246/FREQUENCY MHZ TO DETERMINE WHAT THE
ELEMENT LENGTHS WILL BE.

19 INCH RADIATOR

30LT   BOLT

19 1/4 INCH RADIAL
(1 OF 4)

19 1/4 INCH RADIAL
(1 OF 4)

S0-239 / CONNECTOR

45 DEGREE ANGLE

5½ Wide by 7½ Tall

Figure 4-2: Top view of 2-meter antenna

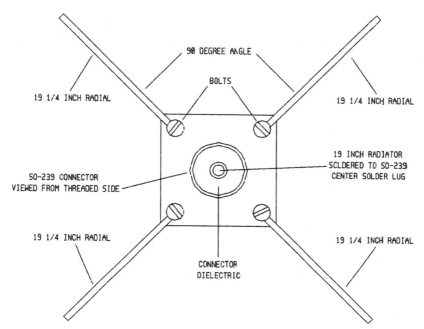

Conclusion

It's a nifty, thrifty ground plane antenna for all occasions. When I first got into HAM radio, I used a similar version on a pole inside my house (while waiting for my base station antenna to arrive). And, I have got to tell you, it worked beautifully.

It didn't have the gain the commercial antenna displayed, but for local 2-meter work, I couldn't complain. I was putting full-strength, full-quieting signals on almost everybody I talked to.

I highly recommend this design and approach. For a portable system you can throw in the truck; it's a winner. I know several HAMs that use these at their base station. Also, as I mentioned, it makes a great scanner antenna.

Building antennas can be a lot of fun, but it can also be frustrating. There is such a science to the subject so, when I find a simple and effective design, I stick with it. Take my word for it, this is just such a design! Enjoy!

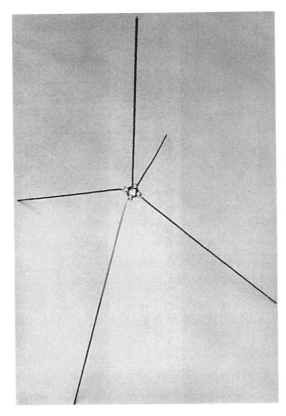

Project 4: The completed 2-meter Ground Plane Antenna with the SO-239 at the center.

Project #5

The Dipole Antenna

Introduction

Here is another simple antenna that has graced amateur radio antenna farms for as long as anyone can remember. By the way, an antenna farm is the place, usually out back somewhere, where a HAM has all his or her antennas. With the more active members of this hobby, they can be quite a sight.

Dipoles are easy to make, very inconspicuous, and CHEAP! The inconspicuous and cheap parts are the nicest of all. Many people don't share our adoration for antennas. They look upon them as eyesores. Sometimes, they are called skyhooks. A skyhook is, to quote Harry Helms AA6FW, "an antenna—especially a large, grotesque one that upsets the neighbors and frightens small children." I couldn't have said it better myself. However, dipoles, for the most part, do not fit in the category of skyhooks. That should keep the neighbors and small children happy.

As I said, they are easy to build and cheap! I always like that cheap part! With a little wire, some PVC pipe, a couple of insulators, and some hardware, you have a very efficient antenna. This is a design many HAMs feel is one of the best ever conceived. On that note, let's take a look at how you, too, can construct the classic dipole antenna. Believe me, this is a really nice way to go when you first get into HF radio.

Construction

Check out the diagram, figure 5-1. A dipole is nothing more that two lengths of wire connected to the feed line at the center. The wire is cut to resonance (length) for whatever band, or frequency, you plan to transmit on, and this is determined by the equation 468/frequency in megahertz.

That formula should look familiar as it is basically the one we used to calculate the lengths for the ground plane antenna. However, this time we want a 1/2 wave antenna instead of a 1/4 wave. Since 1/2 wave is double 1/4 wave, we also double 234 to get the 468 divisor.

Figure 5-1: A typical 1/2 Wave Dipole Antenna

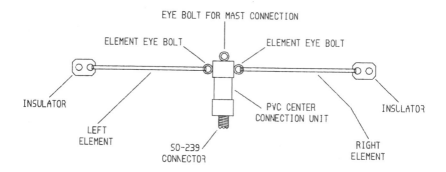

EYE BOLT FOR MAST CONNECTION

ELEMENT EYE BOLT   ELEMENT EYE BOLT

INSULATOR   INSULATOR

LEFT ELEMENT

PVC CENTER CONNECTION UNIT

SO-239 CONNECTOR

RIGHT ELEMENT

HERE IS THE BASIC LAYOUT FOR A DIPOLE ANTENNA. OF CCURSE, THIS DIAGRAM IS NOT TO SCALE. EACH ELEMENT IS CUT TO A SPECIFIC LENGTH AS DETERMINED BY THE EQUATION 468/FREQUENCY IN MEGAHERTZ. THE FREQUENCY COMES FROM THE BAND YOU WILL BE OPERATNG IN. FOR EXAMPLE, IN THE 20 METER BAND, THE FREQUENCY WOULD BE 14.175 MHZ WHICH IS THE CENTER OF THE 14.0 TO 14.35 MEGAHERTZ BAND.

For the prototype, I picked the 10-meter band, as part of this spectrum opens up to technicians when they get their technician plus tickets. Since 10 meters runs from 28.0 to 29.7 megahertz, the elements (wire lengths) are cut for the center of the band, or 28.85 megahertz. Using our handy-dandy formula, that comes out to 16 feet, 3 inches per element (468/28.85). 16 feet is probably close enough, but you want an extra 2 inches on each end to accommodate the installation process.

Now, we need a way to support the elements and at the same time, connect them to our coaxial feed line. That's where the PVC pipe and hardware comes into play. Actually, when I say PVC pipe, I'm referring to a 3½-inch section of 1½-inch (OD) pipe and a couple of end caps. See figure 5-2 for details.

As for the hardware, this consists of three eye bolts, some heavy stranded wire (12-gauge), six ring connectors, two small bolts, and a SO-239 female UHF connector. This, along with the PVC materials, forms the center connection unit for the antenna.

The first step is to drill appropriate holes in the end of one of the caps for the SO-239 connector. It isn't necessary to secure the 239 by all four mounting holes, but you can if you want to. Usually, a large hole for the threaded portion of the connector, and two smaller holes for mounting bolts does the trick.

Next, drill two holes in the side (180 degrees apart) and one in the top of the remaining PVC cap. These are for the eye bolts. It is best to secure the bolts with nuts on both sides of the cap allowing for ring connectors to be placed under the nuts, inside and out, to facilitate connection of the SO-239 and elements.

Figure 5-2: Detailed view of center connection unit

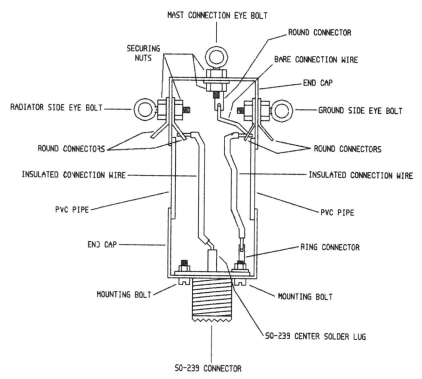

With the bolts and SO-239 in place, assemble the center connection unit by first making the inside hookups. With 12-gauge stranded wire and ring connectors, connect the top and one of the side eye bolts to the mounting flange of the SO-239 connector. The remaining eye bolt is hooked to the center lug of the 239.

You need to leave a little slack in the hook-up wires to allow the unit to be assembled. It is also a good idea to employ PVC cement on the plastic parts when placing them together. That secures the unit. However, test the SO-239 and eye bolts for positive continuity before gluing the unit, as once it is assembled and the glue sets, it isn't coming back apart.

Also, if you use bare wire for these connections, be careful the ground and center conductor do not short together when the unit is assembled. It is probably best to employ insulated wire for the inside work.

With the center connector assembly completed, it is time to turn our attention to the elements. These can be made from a variety of wire types and sizes, as that is not critical. This is one of the reasons the dipole is such a surreptitious antenna. If you utilize small gauge wire of a dark color, it is very hard to see, at least at a distance (no scared kids).

The wire can be insulated or not, depending on what you might have on hand, or can find. Again, this is not a critical factor. However, it is best to use copper wire, as it has superior conductivity to materials like aluminum. Theoretically, that helps your signal get out better.

I would recommend something in the 14- to 16-gauge range, simply to give it more strength. Remember, there is some wind resistance. Very light wire has an annoying habit of breaking, especially if the wind becomes fairly strong.

For the stealth look, 18- or 20-gauge elements might be suitable. They are not as robust, but they might not need to be. That is something your chosen location helps to determine. If you plan to erect the antenna in a fairly sheltered area, wind might not be a problem.

Now that you have decided on the elements, let's hook them to the center connector. If you are using insulated wire, strip about 2 inches of insulation off one end of each wire. Feed about half of this through one of the eye bolts and wrap it around the bare element wire several times to fasten the element.

Next, wrap the heavy stranded wire, that is attached to the ring connector under the bolt, several times around the element winding. With that done, solder these wires together, and use substantial solder. No, I don't mean a whole roll, but be sure this solder joint is well secured.

The last step is to attach the other end of each element to an insulator. This is accomplished in the same wrap-and-solder fashion as before. The insulators are used to fasten the loose ends during installation.

Project 5: Completed 10-Meter Dipole Antenna. This does not show the full length of the antenna elements (wires) but it does show the attachment to the center assembly.

Installation

Speaking of installation, let me talk a little about this. Compared to a lot of antennas, the dipole is easy to situate. All that is needed is a way to attach the center connector unit to a mast (upright) and a way to affix the ends to stable objects or at least semistable objects.

The top eye bolt is ideal for the mast mount. One side of the dipole should be grounded using an earth ground system (copper rod driven into the soil), and the top eye bolt is already hooked to one side of the dipole. Hence, it can also be connected to the earth ground lead.

Since most metal masts are also grounded (or, they should be), conductive attachment of the eye bolt to the mast adds an extra grounding point. However, a conductive connection isn't essential as long as the eye bolt is hooked to a ground line. Thus, cable ties and/or other techniques also work.

Concerning the loose ends, the insulators are now tied to a tree, a fence, another mast, etc., by a nonconductive line. Small rope or marine-style plastic line are good choices. You want to leave a little slack in these lines, and the elements, to keep the wind and/or any tie point movement from snapping your antenna. Also, medium tension springs between the insulators and line can be used to help absorb strain.

When installing your antenna, you have two configuration choices. I covered this in the antenna segment of section two, but let me go over it again. A dipole can use a straight layout or be arranged as an inverted V.

With the straight installation, the elements extend straight out from the support at a 90-degree angle (perpendicularly). The inverted V on the other hand, has the elements slanting downward toward the ground. The angle with respect to the support can be almost anything, however, it is best to keep the overall angle between the two elements at 120 degrees or greater.

Both systems are employed when stringing dipole antennas, but the inverted V tends to take precedence over the straight arrangement. One reason for this is ease of installation. Getting the elements high enough to stay level is not always the simplest of tasks.

Another reason is radiation pattern. The straight dipole tends to be a little on the directional side with the signal extending off the sides of the elements better than off the ends. The inverted V, though, is almost completely omnidirectional in nature. The signal radiates nearly equally in all directions.

Naturally, there are no absolutes with antennas. Each is going to be a little different from the next. So, all of this, to a degree, is theoretical. For the most part, past experience has proven that dipoles function as described for each of those configurations.

Conclusion

There you have it: the dipole. It's one of amateur radio's favorites! As we have seen in this section, the dipole is an easy antenna to build and work with, but it is also a proven performer. I was talking with Dennis, KS4UO, the other day, and he told me that during a contact with Russia, he had a chance to use both his three-element beam and his dipole.

During that QSO, Dennis was receiving a signal level only one dB lower with the dipole than with the beam. So that should give you an idea of how well these simple two-wire antennas work. Two-wire antenna, hum.....is that like a two-horse town? Nah, I don't think so!

Anyway, if you want to put up a quick and simple antenna, without taking out a second mortgage on the house, the dipole should, at the very least, be one of your options. You will have to go a long way to find a better achiever at this price!

Project #6

A Cubical Quad Beam Antenna

Introduction

So far, we have built only omnidirectional antennas, but here is one that is a very efficient beam antenna. Known as the "Cubical Quad", or just "Quad", this falls into a category of directional antennas that are both inexpensive and easy to build. Additionally, they don't require a whole lot of tuning to get them running.

What we are going to construct is a two element quad for 2 meters. This one comes to us from Pete KF4QOE, and does a remarkable job for its size. That's something else likeable about this gem. It is quite small, thus very portable.

Built from easily obtainable materials, the finished antenna shouldn't put you back more than $20 and that is if you buy everything new. Also, I haven't been to a HAMfest yet that didn't have someone selling most of the parts for antennas like this one, which usually saves some coins. So, keep an ear open for a HAMfest in your area. With that said, let's take a look at how to construct this antenna. It really is very easy, and the finished product is a delight!

Construction

Figure 6-1 gives an overview (diagram) of how the two-element quad is fabricated. You'll note it consists primarily of two box-shaped elements on a boom. Hence, the cubical part of its name. This design has been around for just about as long as HAM radio itself, and is a proven winner.

Figure 6-1: 2-Meter Cubical Quad Antenna

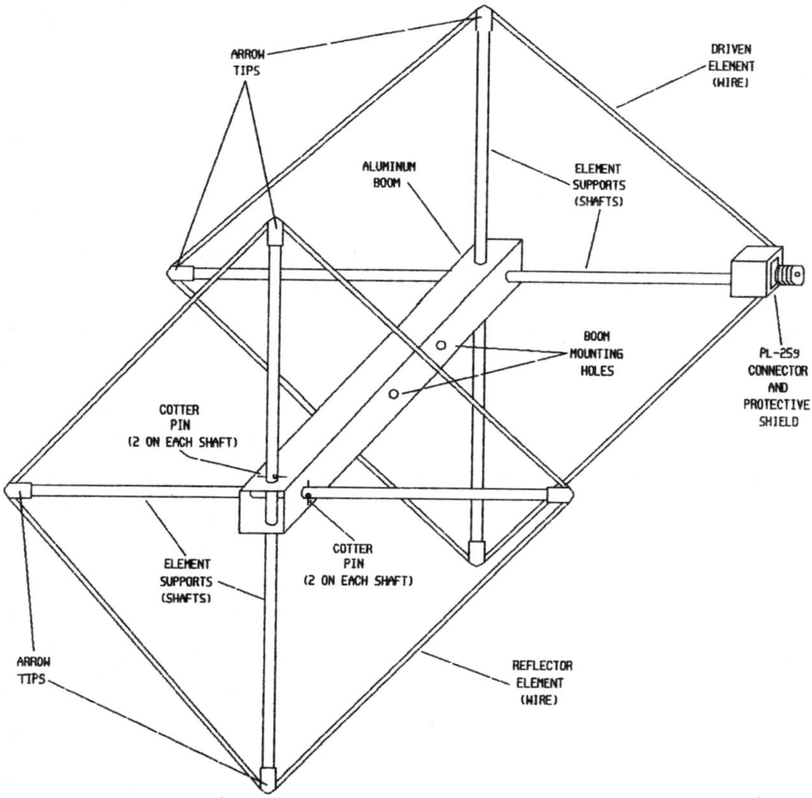

ARROW TIPS

DRIVEN ELEMENT (WIRE)

ALUMINUM BOOM

ELEMENT SUPPORTS (SHAFTS)

BOOM MOUNTING HOLES

PL-259 CONNECTOR AND PROTECTIVE SHIELD

COTTER PIN (2 ON EACH SHAFT)

ELEMENT SUPPORTS (SHAFTS)

COTTER PIN (2 ON EACH SHAFT)

ARROW TIPS

REFLECTOR ELEMENT (WIRE)

I'm getting off track. To start with, you need four sections of 1/4-inch diameter fiberglass rod, each at least 33 inches long. One great source for this is a sporting goods store, as the prototype antenna utilized arrow shafts for this.

While you're at the sporting goods store, pick up eight arrow tips. This is the little plastic piece on the end of the arrow that you slip the bow string into, not the sharp end! The tip holds the arrow so that when you pull the string back, then release it, the arrow is propelled forward. At least, that the way it is supposed to work. I don't know, with my adventures in archery, it hasn't always worked like that. But, that's another story. Let's get back to our antenna.

The arrow tips are used to support the wire elements of the quad. So, you also need one section of 14-gauge insulated solid wire 86 inches long and another 81 inches long. The reason for the different lengths becomes clear as we go along. The longer section (reflector) has to be soldered together to form a circle, then the solder joint should be covered (protected) with either electrical tape, or better, heat-shrink tubing. That is the stuff you slip over a solder joint then heat. When heated, it shrinks around the joint forming a tight seal. Actually, its name is a misnomer. In reality, this is plastic with a memory that has been stretched to roughly twice its normal diameter. When the heat is applied, it returns to (remembers) its original shape. Just a little trivia, there!

A third item you need is a 16-inch section of 1-inch square aluminum tubing (PVC pipe can also be used, but I think square aluminum is a better choice). Most of this is used for the boom, with a short 1-inch section salvaged to construct a protective housing for the antenna connector.

With these materials in hand, let's start construction on the quad. There is some other stuff required, but we get in to that as we go along. The first step is to cut the aluminum tube to 13¼ inches in length. That creates the boom, or center support. Next, starting at about 3/4 of an inch from the end, drill 1/4-inch holes in the boom for the fiberglass shafts, and a couple of holes in the center to oblige a U bolt for mounting. The size and spacing of these holes depend on the U bolt you use.

The holes on the ends have to accommodate pairs of vertical and horizontal shafts with just enough clearance so the shafts do not bind against each other when they go through the boom. It doesn't matter whether you start with the horizontal or vertical supports, but you must be consistent.

That is, if the vertical is on the outside of one end, the other vertical must be on the inside of the opposite end. The reason for this is that you need 11½ inches between each set of vertical supports and the same for the horizontal supports (see figure 6-1 for details).

Next, comes the shafts. They need to be cut to 31 inches for one pair and 29 inches for the other pair. It is best to cut them slightly long, and trim if necessary, as it is their length that keeps the wire elements taut.

This is important! If the wire elements display any slack, the standing wave ratio (SWR) suffers. You want the wire to stretch tightly around the perimeter of the quad frame. That is one of the reasons fiberglass is suggested. There is going to be substantial tension on these cross-supports, and the fiberglass arrow shafts are far less likely to snap under that tension.

Once you have positioned the shafts, drill small holes in them just above and just below the boom. Then, place small cotter pins in these holes to prevent the shafts from sliding. When all four cross supports are secured in place, put the arrow tips on the ends of each. The frame is ready for the wire elements.

Since the longer element (reflector) has been soldered together, start with it. It should now be in a somewhat circular shape, and you need to slip this into one of the arrow tips, then the next, and the next and the last one until the element is in place around the frame. It is now a square shape, as opposed to round.

This process requires flexing the shafts, especially the last one, in order to get the wire in all four tips. Here again, the fiberglass is much more forgiving than other materials.

For the shorter (driven) element, an SO-239 antenna connector is attached to the ends. One end is soldered to the connector's center tab, and the other to ground. The ground connection can be done with a small bolt. A 1 inch section of the remaining aluminum tube is used to protect the connector from the weather. This is placed over the wire and brought down to the connector before the element is installed on the arrow tips.

Now, the wire is slipped into each tip around the perimeter of the frame, with the connector/aluminum piece lining up with one of the arrow shafts. This can be secured to the shaft with a couple of cable ties. Pete recommends filling the aluminum shield with hard-set epoxy cement and allowing it to dry (set up).

With all that done, you are just about done. However, there is one last step that is extremely important for correct operation of your quad antenna. This is the matching network that adjusts the rather high antenna impedance to around 50 ohms. That, of course, is what the radio wants to see.

Actually, this is pretty simple. You need 12½ to 13 inches of RG-59 coaxial cable, two PL-259 connectors and a barrel connector. The PL-259s are installed on each end of the coax, and the barrel connector (two SO-239s back to back) is used as a female antenna connector by hooking it to one of the 259s. The other PL-259 is attached to the SO-239 on the antenna.

Project 6: Completed Cubical Quad Antenna on a portable mast. Note the matching harness at the right.

This arrangement balances the impedance of the antenna to a value that your radio(s) can use. Without it, a bad mismatch exists, and that definitely inhibits the antenna's performance.

Now, all that is left to do is mount the quad to a vertical support of some sort. This could be a length of metal or plastic pipe, or you can mount it to an existing mast or tower. Once in place, the 2-meter cubical quad is ready for action.

Test and Application

In reality, there isn't much to do in the way of testing this antenna. If you place an ohm meter across the connector, you see a dead short. But, believe it or not, a dead short does not exist. Due to the nature of this design, an applied RF signal radiates nicely off the driven element. However, to look at it, you would swear there is going to be trouble.

This is a phenomena encountered often with antenna designs. J-poles and UHF Loops both appear to create that same dead short, but in operation, both work like a champ. As I said before in this text, antennas are black magic.

As for employment, the antenna doesn't do windows! No, just kidding! Just a little joke! Geeez....you don't have to get hostile! Anyway, where was I? Oh yeah, operation. This part is great, especially if you're looking for a highly sensitive, and selective, antenna.

Unlike omnidirectional antennas, beams have the ability to condense incoming signals, thus highly increasing their strength. Our little quad is no exception. Hook it up to your radio, and begin rotating it. Soon, you begin receiving very distinct signals from other radio stations.

In whichever direction you have the reflector element pointed, that is the direction of the station. The more precisely you aim the quad, the stronger the signal is. This also applies to your transmitted signals.

Once you zero in on the other station (find its location), your transmissions become many times stronger than if you were using an omni. This can make a big difference when distance is a factor. You are able to communicate with stations much farther away with this antenna.

So, all there is to it is to rotate the quad until you get the clearest signal. At that point, you are lined up with said station. Let the QSO party begin!

Conclusion

We looked at two other antennas in this book, and a fourth is still to come. But, I think Pete's Quad is one of the best projects you'll undertake. Not only is it easy to construct and CHEAP (There it is again, one of my favorite things!), but the performance is truly incredible.

If you are into DXing (contacting distant stations), or like to get out mobile and chase foxes, this antenna is for you. Its ability to pinpoint a signal, and also amplify it, makes the quad exemplary for both activities.

So, go ahead! Give this one a shot. I doubt you will regret such a decision, as you WILL find a use for this beauty. Even if it is at some point down the road of HAM radio. What you learn along that byway makes the effort worthwhile. I know I stress the educational part of all this a lot, but that too is part of amateur radio.

Project #7

A Multiband Vertical Antenna

Introduction

Here is an antenna that comes to us from Dennis KS4UO, and handles the 10-, 12-, 15-, 17- and 20-meter bands. It is a little more complicated than the first three, but not much! If you are ready to really jump into HF radio, this antenna just might be your ticket.

Dennis got the idea for this multiband vertical when someone was going to throw away a few tent poles and asked him if he wanted them. Being a good HAM (a scrounger), he said sure! This multiband design was born!

One of the really nice aspects of this system is its portability. The poles easily come apart, and that makes for one handy antenna for field use. However, it also makes a dynamite base antenna on the aforementioned bands.

With the addition of some wire to the top element, its performance could be extended into the 40-and 80-meter bands. That is something Dennis is working on as of this writing.

So, now that you know a little about this antenna, let's take a closer look. This gives you the opportunity to better understand its operation, and also explains how you can build one of your very own.

Theory

The principal behind this design is to add sections to the antenna as the operating frequency goes down. As I'm sure you have noticed by now, the lower the frequency, the longer the antenna elements.

With ground planes and quads, the elements themselves have to be increased, while dipoles are often constructed with sections of wire separated by insulators. Jumpers then connect the sections, as needed, to obtain the proper element length.

For this vertical, the tent poles are moved in and out of the system to arrive at the proper element size. The top section actually fits inside the section directly below it, and this allows for fine tuning the antenna.

That fine tuning amounts to releasing a hose clamp that keeps pressure on the outside element and sliding the smaller top element up and down until the minimum standing wave ratio (SWR) is achieved. Naturally, this process changes the total length of the radiator by a small amount. That is the key!

Since this vertical is a specialized ground plane, we need some radials. Dennis has those for us. Boy, does he have those; 30 in all! Actually, these lay on the ground below the vertical element, which is a common practice in amateur radio especially with HF antennas.

With the portable version, the radials are simply extended out away from the main radiator and spread in a 360-degree fan arrangement. This works well, but you do want to be careful not to get tangled up in all that wire.

For a permanent installation (base station), it would be best to bury the radials a few inches underground. That eliminates the tangling problem in the radial system and also gives slightly better performance. Again, a fan-shaped arrangement would be advised. This provides more evenly distributed grounding on the system, which usually translates into a better radiation pattern.

That is the basic theory behind this antenna. Next, let's look at how to construct it. With some commonly available materials, and a few tools, you can have one of these up and operating at your station.

Construction

As mentioned, Dennis constructed this system using several orphaned tent poles, but any type of aluminum tubing or pipe works. They need to fit together (have insert flanges) if you plan on portable employment. The top section has to be small enough to slide in and out of the other sections.

The maximum length of the antenna, for the 20-meter band, is 14.08 feet and the shortest height, for 10 meters, is 6 feet. At first glance, two 6-foot sections and a slightly smaller 3- to 4-foot section would seem ideal for the job. But, due to the radiator lengths of the intermediate bands, four 4-foot sections are required. This makes the antenna easier to transport, however.

To start with, the base section and the section next to the top need a slot, about an inch deep, cut across the end. This allows the tubing/pole to compress slightly when a

Project 7: A close-up of the Multiband Vertical wiring harness. Note the radials leading down to the ground.

hose clamp is tightened to hold the top section in place. A hacksaw does a fine job in cutting these slots.

The bottom section, in addition to being part of the radia-tor, also supports the radials. These are connected in bundles of five radials each and are cut to 1/4 wave length for each of the bands previously mentioned. For example, a bundle for 12 meters would be five 9-1/3-foot leads. These can be just about any gauge wire, but 16- to 20-gauge probably works the best.

Figure 7-1: Multiband Vertical Antenna

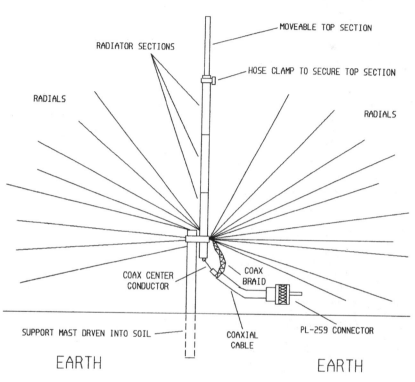

HERE IA A DIAGRAM OF THE GENERAL LAYOUT OF THE MULTI-BAND VERTICAL
ANTENNA. THIS IS NOT TO SCALE, BUT IT SHOULD GIVE AN IDEA OF HOW THE
SYSTEM IS ARRANGED. NOTE: FOR CLARITY, NOT ALL RADIALS ARE SHOWN.

Table 7-1 provides the specifications for both the radiator and the radials. This table covers the five bands in which the antenna is designed to operate. Notice that Dennis divided the 10-meter band in half, and thus there are 10 radials for 10 meters because that band is rather large to start with. Hence, for the five bands you have 30 radials instead of 25.

The feed point for the system is located 25 inches above ground on the bottom section. This is the point where the coaxial cable connects to the radiator and radials. As with most antennas, the center conductor of the cable is the radiator feed, while the coaxial shield or braid is the ground (radials).

When hooking up the cable, be very sure the radials do not make electrical contact with the vertical radiator (use heat-shrink tubing or electrical tape around the radial solder connection). Electrical contact between the two would result in a dead short and possible damage to any radio connected to the antenna. Again, I KNOW you DON'T want that!

With the antenna constructed, it can be mounted to a pole driven into the earth. This furnishes a more positive ground for the system, but be sure the feed point is at least 25 inches above the soil. You might encounter interaction between the radiator and ground if it is closer than 25 inches.

That pretty much covers construction of the antenna. Let's now look at how to set up, tune, and use this beauty.

Table 7-1: Radiator and Radial Dimensions

Radiator

Band	Frequency	Size
10-Meters	28.0 - 29.7 MHz	72" / 6 ft.
12-Meters	24.89 - 24.99 MHz	89" / 7.4 ft.
15-Meters	21.0 - 21.45 MHz	112" / 9.33 ft.
17-Meters	18.07 - 18.17 MHz	135" / 11.25 ft.
20-Meters	14.0 - 14.35 MHz	169" / 14 ft.

Radials

Band	Frequency	Length
10-Meters	29.600 MHz	7.9 ft.
10-Meters	28.500 MHz	8.2 ft.
12-Meters	24.940 MHz	9.4 ft.
15-Meters	21.225 MHz	11 ft.
17-Meters	18.110 MHz	12.92 ft.
20-Meters	14.175 MHz	16.5 ft.

The radials come in groups of five each. Each ground represents one of the bands listed. All radial are electrically connected together, so you are actually using all 30 radials for each radiator length. This is acceptable because the more radials the better.

Tuning and Application

The first step in getting the multiband vertical ready for action is to decide what band you want to work. That determines the number of poles needed for the radiator. Let's say the 17-meter band is your choice. The radiator length for 17 meters is 11 feet, 3 inches. That means you need two of the 4 foot poles and the smaller top section to configure the vertical for 17 meters. Assemble the two lower sections, then insert the smaller section in the upper pole (the one with a slot). Position the top section so that 3 feet, 3 inches are exposed and lock it in place with a hose clamp.

That comprises the initial setup. However, for optimum performance, you want to fine-tune the radiator. This is where your SWR meter, or antenna analyzer, comes into play. With the meter/analyzer in line with the antenna and radio (between the two), transmit a test signal and check the SWR reading. If the reading is in the 1.3:1 range, or lower, then you probably have it tuned about as well as you're going to get it tuned. If however, the reading is higher than 1.3:1, adjusting the antenna should improve the SWR.

That adjustment is done by moving the top section up or down. Start by raising the element (top section) by 1/4 inch. Then, send another test signal and again check the SWR. If the reading goes down, you're on the right track, so try raising it another 1/4 inch. At some point, the SWR may be going up again, and that indicates that the previous position was the best possible setting for the top section. If, however, the first move garners a higher SWR, you are going in the wrong direction.

In this situation, move the element down by 1/2 inch (1/4 inch to return to the starting point and an additional 1/4 inch for the new setting). The SWR should go down, at least some. If not, try another 1/4-inch movement. If nothing changes, then the antenna may be as good as it is going to get. However, you should see some fluctuation in SWR as you change the top section's location.

Project 7: The completed and temporarily installed Multi-band Vertical. Dennis KS4UO checks the SWR.

Naturally, if the SWR goes up when you move the element down, as it did when you raised it, the original position was correct. It is unlikely to hit this condition on the first try, but it can happen. In fact, it happened to me with my very first base station antenna. I set the moveable element to the manufacturer's recommended setting, and tried it. The SWR meter barely moved. So, I said to myself, "Self, why don't you try another setting to see if you can get the antenna flat (1:1 SWR)?" I agreed with myself and did just that. After several tries though, I came back to the original position, as that was the best.

Anyway, once you find the finest possible SWR, your multi-band vertical antenna is ready for use. At this point, take the meter out of the line, or you could leave it in if you like, and begin roaming your chosen band for activity. It shouldn't be long before you're engaged in a lively QSO with another station.

Conclusion

Okay! Have some fun with this one. Unlike the other antennas we have built, this one gives you some room to play. That is not only educational (there's that word again) but also allows for optimum performance. In the end, I suspect you will find this design does an excellent job.

KS4UO is in the process of fully evaluating his prototype's performance, but so far, he is quite happy with the preliminary results. By all appearances, this one seems to be a top achiever. But then, its design springs from the sound foundation of ground planes, hence, it should be good.

Who knows? With your portable version in hand, you may well be the star, hit, and champion of your club's next Field Day! A top performer at Field Day is worth its weight in gold. Believe me, I know of what I speak from painful ex-perience!

Project #8

The 40-Meter QRP Station

Introduction

Okay gang! Here is the kit! Actually, this project uses two kits from Vectronics of Starkville, Mississippi, that are easy to build, and even easier to tune up. One is a transmitter and the other a receiver. Both are geared for CW operation on the 40-meter amateur band (7.00 to 7.30 megahertz). That makes them excellent for honing your continuous wave (CW) skills.

Of course, as a Technician you are not allowed on this band, but with your Technician Plus ticket, 7.10 to 7.15 megahertz is available for CW contacts. As stated, that type of operation helps perfect CW proficiency and allows you to upgrade your license.

This system is also a QRP station (low power). The transmitter puts out a mere 1 watt, so contacts on this rig are really challenging. However, you'll be amazed at how many you make! In the 40-meter band, signals can travel remarkable distances with this kind of power.

So, enough of the big pitch! Let's delve into the worlds of 40-meter QRP, and kit building. Both benefit your knowledge of amateur radio and provide a whole heap of fun along the way.

Transmitter Theory

The transmitter is a very simple arrangement that does not require any tuning. When the key (transmit switch) is depressed, it activates a transistor that, in turn, powers the oscillator stage. This is a Colpitts oscillator that determines operating frequency and also supplies the Variable Frequency Oscillator (VXO) signal.

The transmit frequency is determined by a crystal (rock bound), and the unit comes with one such crystal for 7.040 megahertz. A second crystal can be selected via a switch. This is good, as 7.040 MHz is a popular QRP frequency, but it can't be used with a Tech Plus license (remember, a Tech Plus is limited to 7.10 to 7.15 MHz).

So, the second crystal needs to fall within that frequency range (see source list for crystal dealers). That is, until you get your General class license, of course. Then the entire band is open to you.

Anyway, back to the theory. The same transistor that activates the oscillator also powers a third transistor that acts as a driver for the power amplifier (PA). The term driver here means this transistor develops enough output to turn on the fourth transistor, or power amplifier.

While in the ON mode, the PA has voltage on it at all times, but it only draws current when the key is depressed. In that fashion, it is, in effect, OFF until you key the rig.

Using a couple of diodes, a transmit/receive (T/R) switch is present that automatically disconnects the antenna from the receiver when the transmitter is operational. This keeps the transmitter from swamping (overwhelming) the receiver.

And, a pi-style low-pass filter handles harmonic suppression. This keeps the transmitter from sending strong signals on bands it is not supposed to be sending signals on. For example, 20 meters, or the second harmonic of 40 meters. The FCC really appreciates this part of the circuit.

That is about all there is to it. It may sound a little complicated, but in contrast to many other radios, this is a simple, but effective, transmitter design.

Receiver Theory

The receiver utilized direct conversion to eliminate the need for inter-frequency (IF) sections. This provides a much less complex design, but does dampen the sensitivity a bit. However, this unit is rated a 3-microvolt sensitivity (the receiver detects a signal as low as 3-microvolts in strength), so that dampening is minimal.

A potentiometer is used as a gain control and also helps keep strong nearby stations from overloading the front end (initial input section of the receiver). From there, a variable capacitor and a few coils (inductors) handle the antenna impedance matching and bandpass filtering.

The resulting signal is sent to an NE602 integrated circuit (IC) that acts as the mixer and local oscillator (LO). The mixer is configured as a Gilbert cell which furnishes low noise and excellent gain. The LO is loaded heavily with capacitance to keep frequency drift to a minimum.

During reception, the local oscillator is tuned to either the single sideband (SSB) carrier, or a range of 300 to 800 hertz for CW. This produces an audio signal for amplification. Speaking of amplification, that task is accomplished by another integrated circuit, an LM386 low power audio amp.

The LM386 is an electronics hobby classic that incorporates, on a single chip (IC), just about everything you need for a fully functional audio amplifier. By coupling the LO to the 386 input, headphones, or even a small speaker, can be used to hear the incoming signals, as this chip yields about 500 milliwatts of audio power.

Stability for the circuit is achieved by using a third IC, an LM78L05 linear voltage regulator. This chip keeps the voltage steady regardless of the input voltage, and helps maintain rock-solid performance.

Incidentally, the receiver can be configured for the 80-, 75-, 30-, and 20-meter bands, in addition to 40 meters. That provides some real versatility, and the manual that comes with your kit gives details on this. It also receives AM, SSB and RTTY in addition to the CW mode.

That about covers the technical side of both the transmitter and the receiver. All of this is good information to have, but let's get on to the fun part: building your 40-meter QRP station.

Construction

Both of these are very nice kits. That is, they are well made, have quality components and have very good instructions on how to assemble them. That, of course, makes them delightful to build.

For each, everything you need is included, and all that is necessary is to follow the step-by-step directions in the manual. You can't miss! The receiver does require winding a coil, but this is covered in great detail, and presents no problem.

Be careful with your soldering technique (no cold or sloppy joints) and watch the polarity and/or orientation on any applicable component. If you do that, you shouldn't have any problems completing these kits. The nicest part is, they actually work when you're finished.

I must say I was truly amazed at the lack of pitfalls that often accompany kits. You know what I mean with inadequate instructions, vague terminology, misidentified parts and the like. With these kits, I found none of that.

Both the transmitter and receiver have well-constructed and marked printed circuit boards (PCBs), which helps to speed up the assembly process. The provided components are the right size for the holes drilled in the boards. This last one might sound like nit-picking, but when you have built as many kits as I have, this is important.

As for the final touch — the enclosure(s) — you can purchase them from the manufacturer or scrounge up your own. I used a case that began life as a satellite receiver decoder, or something like that, because it was the right size and shape. Virtually any box you have, or can find, that fits the two boards should do nicely.

With my system, I drilled the appropriate hole to match the various control functions of the units (tuning, On-Off switches, crystal select, etc.) in the front panel. I also drilled holes in the back panel to accommodate the headphones, the antenna and power inputs.

These panels were then labeled with dry transfer lettering, and a light coat of clear enamel was sprayed on to protect the delicate lettering. Keep the protective coat light, as the enamel solvent melts some dry transfer material.

Once the case is ready, the two boards are secured to the bottom with bolts and stand-offs. The final wiring is done, and the station is ready for the receiver to be tuned. Then, the whole works needs testing.

All in all, you are in for a highly pleasant experience with these gems. I worked slowly and only spent about four hours on both projects combined. With this project, your 40-meter QRP station is up and running that much faster.

Tune-Up and Operation

As previously stated, the transmitter doesn't require tuning, which is nice. The receiver, however, does demand a little bit of attention to make it ready for the 40-meter band. This procedure involves aligning the local oscillator and can been done in one of three ways.

Figure 8-1: 40-Meter QRP "Dummy Load"

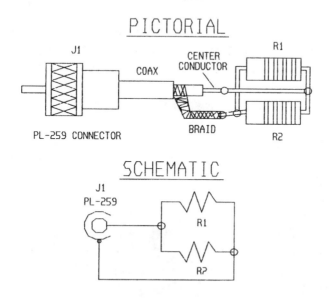

PICTORIAL

SCHEMATIC

PARTS LIST:

R1 AND R2: 100 OHM, 2 WATT RESISTORS

J1: PL-259 COAXIAL CABLE CONNECTOR

The methods include using another receiver, using a signal generator or using the companion transmitter. Since we have the companion transmitter, that is the way to go. You need a dummy load for this, and figure 8-1 illustrates how to make one.

With the dummy load in place (connected to the antenna), and the receiver functional (headphones/speaker connected to audio output), set the receive gain fully clockwise, and the tuning control to a position that is roughly the frequency you want to receive (7.040 MHz). Now, with a small blade-style tuning tool, bring the core of coil L1 to the top of its form. A nonmetallic tool is preferred.

Next, with the transmitter sending a signal into the dummy load, adjust L1 for the strongest receive signal. Do this slowly and in small increments.

Project 8: The two boards that make up the 40-meter QRP Station. Each is a kit with the transmitter in front and the receiver behind.

That procedure sets the tuning range, and now the RF input stage must be fine-tuned for peak performance. This involves setting variable capacitor C18.

Again with the tuning tool, adjust C18 for the loudest audio. The gain potentiometer probably will have to be backed off (set counterclockwise) to prevent overloading the receiver. When you have the loudest possible signal, you have the RF input peaked. Basically, that is all there is to tuning the receiver. Trust me on this, that is an extremely simple tuning procedure for a receiver. The direct conversion design has a lot to do with the simplicity.

Now that you are all tuned up, your QRP station is ready for service. For on-air operation, you need to connect a suitable antenna (perhaps a dipole) to the system's antenna input. The feed line needs to be 50-ohm coaxial cable (RG-8, RG-58, etc.).

Also, a jumper cable has to be run between the receiver and transmitter for the automatic Transmit/Receive (T/R) switching circuit. However, the kits do incorporate RCA-style jacks for this purpose, so it is a simple matter.

Merely run a short length of RG-58, with RCA plugs on each end, from the jack marked ANTENNA on the receiver board to the jack marked RX ANT OUT on the transmitter board. The regular antenna feed line connects to the transmitter jack marked ANT. This interconnects the two boards for proper operation.

Regarding the front panel controls, the receiver has an On-Off power switch, tuning and gain controls. These are fairly

Project 8: An inside view of the 40-meter separate transmitter/receiver QRP station. The receiver is to the left and transmitter on the right.

self-explanatory, with the gain control adjusting the receiver's volume. The transmitter has switches for power and crystal select, and the VXO tuning dial. That last one determines the tone (frequency) of the CW signal that is sent. Also, a jack is present for the key.

So, this system is no more difficult to operate than it is to build. With a little practice, you can have it mastered in no time at all.

Conclusion

In this day and age of transceivers, our QRP station represents a blast from the past. That is, both a separate transmitter and receiver are employed to fabricate the station. Not much of that is seen anymore, especially concerning the newer equipment on today's market.

That is, in some respects, a pity, as it is great fun to use a system that does it the old way. Also, if you just want to monitor 40-meters, you can turn the transmitter off. You can't do that with a transceiver!

Seriously, this endeavor is a superb learning experience. If you have an interest in the technical side of amateur radio, there is NO better teacher than hands-on assembly of your own equipment.

That is not to say that all your gear has to be built with your own two hands. Not at all! There is a veritable cosmos of excellent radios, as well as other equipment, available that would be, for the most part, prohibitive to build yourself. I mean, the complexity of much of the present day stuff makes homebrew next to impossible.

However, on the other side of that coin comes the gear like this station. Even though we love all the whistles and bells found on the newer equipment, HAM radio doesn't have to be complicated. The pride associated with working with a station you have constructed from scratch often makes that approach appealing.

So, if you haven't tried the kits, or building from a text/ magazine article, you may be missing one of the real delights of amateur radio. At least, you will find out a little about how all these magical radios work. That alone makes it worthwhile. Have fun and enjoy the thrill of homebrew!

Project #9

Simple SWR Meter for 1 to 300 megahertz

Introduction

One of the most useful gadgets you can put in your radio shack is a Standing Wave Ratio (SWR) meter. We have talked at length about the importance of a good antenna system, and one of the primary methods of evaluating your antenna is determining the standing wave ratio. To put it simply, SWR is the amount of signal that is not getting to the antenna. In effect, that part of the signal remains "standing" on the feed line, and is lost energy. Furthermore, it can come back into the radio and do egregious harm to the radio.

So, with any antenna system existing or new, it is a darn good idea to keep track of the SWR. An especially nice aspect of this project is that the meter is simple and inexpensive to build. Additionally, it can remain in the feed line for constant monitoring without any effect on the radiated signal. With all that going for it, how can you not build this gem?

Theory

As stated above, when you transmit, not all of the radio's signal is going to radiate from the antenna. A small portion will remain on the feed line, and that is, indeed, lost signal (energy). Naturally, you want to get as much of the power out into the air as possible, but there are demons out there that want it the other way. They are nasty little creatures, with names like "impedance" and "mismatch" that exist only to make your life miserable. And, in the name of serious "HAMing," it is our duty to drive these beasts from our midst.

Well, maybe I'm being a little hard on them. On second thought, NO, I'm not! They are out to get our signals and ruin our radios, so down with the infidels.

Alright, I have regained my composure, so let's continue to discuss how this SWR meter works. After all, that was what I intended to do in the first place. I do get sidetracked at times, but in this case, it may be for best. It is good to be aware of those monsters!

Looking at Figure 9-1, the schematic diagram, you will see the meter is comprised of two sections. The first is a simple "path" through the unit from the input (transmitter) coaxial connector to the output (antenna) coaxial connector. This is why the meter can remain "in-line" at all times without affecting the signal.

The second section does the signal monitoring. Here, two smaller copper traces on the printed circuit board (PCB) run parallel to the main trace between the two connectors and inductively detect current moving on the main trace. This is accomplished through two individual "passive de-tecting" circuits comprised of C1, D1, and R1, and C2, D2, and R2. Potentiometer R3 sets the calibration level, and, of course, meter M1 provides the readings.

Now, for the meter to be of use, two different measure-ments must be made of that detected current. First, you

Figure 9-1: SWR Meter Schematic

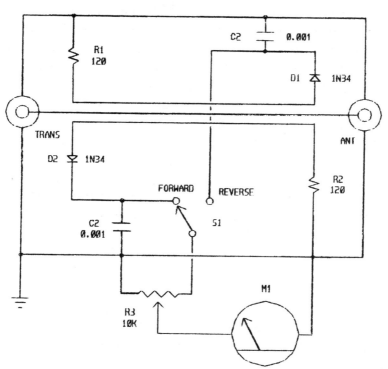

must apply the current to a circuit that will read its "forward" movement. This is done to calibrate the meter for various transmitters and power ranges. Once you have the meter calibrated, you will then be able to measure the amount of signal standing on the line (reverse).

Switch S1 facilitates which measurement you are taking, and the two identical detector circuits are used to observe the current flow. The circuit closest to the transmitter connector detects the forward movement, while the circuit closest to the antenna connector is used for the reverse measurement.

A simple diode/capacitor arrangement makes up the passive detectors that monitor the signal, and the result is sent through a potentiometer (R3) to a 250-microvolt analog meter movement (M1). As stated previously, R3 is the calibration control. The meter is marked appropriately (Figure 9-2) to read the actual SWR when S1 is in the reverse position. A full-scale reading is the calibration point when the switch is in that position. And, that is all there is to it! Let's next take a look at how to build this device.

Figure 9-2: Meter face diagram

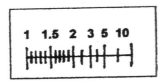

Construction

Probably the most critical aspect of this meter is the printed circuit board (PCB). An exact-size template is provided as Figure 9-3; this will allow you to reproduce the PCB for your unit. As can be seen, it is a fairly small board that accommodates a fairly small and convenient meter.

Looking at the PCB pattern and photograph 9-1 you will notice the main trace running through the middle. In this configuration, I have bent it first down, then up, then back down to reduce the overall length of the PCB, hence, the meter itself. This trace needs to measure 3½ inches in total length (from connector center conductor to connector center conductor), but if you want a longer meter, it could be configured in a straight line.

Figure 9-3: SWR Meter PCB pattern

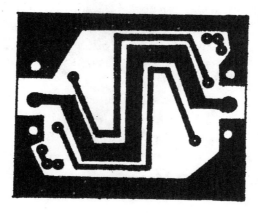

In either case, the two parallel side traces need to be separated from the main trace by 1/16 of an inch. This distance will furnish the induction needed to transfer enough of the transmitter's signal to the detection sections of the meter. Aside from that, the circuit is really noncritical.

I elected to use the surface-mount technique to assemble the meter, as seen in Photo 9-1. The primary reason for this approach is to allow the backside of the PCB to remain a solid copper sheet. This forms a ground plane that will help prevent extraneous RF energy (such as intermodulation) from getting into the meter and messing up your reading. The reverse side of the board is illustrated in Photo 9-2, and it is necessary to grind away a small amount of the copper around the holes where the two connector's center conductor pins pass through. This is done to prevent a direct short between the connector's center pin and the outer ground shield. Trust me on this; your transmitter will not take kindly to a direct short here! On a final note, at two spots on the board (upper and lower edges), I

Photo 9-1: Close-up view of the backside of the SWR meter board. Note it retains the copper layer as an interference shield.

Photo 9-2: Close-up view of the front side of the SWR meter board. Note the copper traces and surface-mount construction.

inserted short pieces of lead wire through the board and soldered them to each side. This is done merely to insure a good and even ground plane on both sides of the PCB.

Once you have the PCB assembled, it can be installed in a small plastic or metal project box. All that is necessary is to cut a hole for the face of the meter movement, one for the calibrate/SWR switch, and drill a hole for the potentiometer in the front panel. For the back of the box, drill two holes to accommodate the SO-239 connectors. A little labeling will indicate what the various controls do.

Operation

There is nothing to using this device. Simply connect your transmitter/transceiver to the input connector (transmitter) and the antenna to the output connector (antenna). Find a clear frequency, set switch S1 to forward and key up. Next, adjust potentiometer R3 until the meter's needle reaches full scale (all the way to the 10 position). Now, release the transmit switch, and move S1 to the reverse position. Again, key up, and the meter will automatically indicate the standing wave ratio.

As stated, the unit can be left in the feed line for constant SWR monitoring or moved between several different transmitters for the occasional SWR test. Not much too it at all!

Conclusion

All right! You did it! Now you have one handy-dandy, nifty-thrifty standing wave ratio meter that will measure SWR in the 1- to 300-megahertz range. That means everything from 160 meters to 1.25 centimeters, and that is a lot of the amateur radio spectrum!

Standing wave ratio (SWR) is an important measurement to watch! Not only will your equipment function far better when it is low, but a high SWR can be downright hazardous to your radio's health! If enough of the signal backs up into the transmitter's final RF amplifier, the result could well be a lot of smoke and a dead radio. That event will likely result in some intense screaming, cursing, and wall-kicking on your part, so do keep an eye on the ol' SWR, as I hate to see a grown HAM cry!

Fortunately, most modern day transceivers utilize detection circuits that help prevent this from happening. As the SWR increases, the radio actually reduces the output power to protect again damage. But, for many older rigs, and most homebrew gear, that protection doesn't exist, and the meter becomes your only defense against high SWR.

Hence, this is a good one to construct! Have fun with it!

Table 9-1

Simple SWR Meter Parts List

SEMICONDUCTORS

D1, D2- 1N34 Germanium Diodes

RESISTORS

R1, R2- 120 Ohm ¼ Watt Resistors
R3- 10,000 Ohm Panel Potentiometer

CAPACITORS

C1, C2- 0.001 Microfarad Disk Capacitors

OTHER COMPONENTS

M1- 250 Microamp Analog Meter Movement
S1- SPDT Slide, Toggle, Etc. Switch
Connectors- SO-239 Female Coaxial Connector (50 Ohms)

Project #10

Tiny 40-Meter CW Transmitter

Introduction

One of the best bands to work low-power continuous wave (CW) is 40 meters! In that essence, here is a very small transmitter that will produce between 1 and 3 watts of power. This puts it in the low-power, or QRP, operation category, and its 1½- by 1¾- by ¾-inch size makes it ideal for portable operation.

Additionally, the component count is minimal (10 parts), the toroid coil is easy to wind and the overall layout is anything but critical! I mean, what else can you ask for? You can knock this one out in a single evening!

However, don't let the small size and part count fool you. When 40 is open, this little gem will get you on the air big time! With a good antenna and receiver, you will be able to work stations at astounding distances when conditions are right. So, let's take a closer look at our 40-meter QRP transmitter.

Theory

Basically, this device is nothing more than an oscillator that is turned ON and OFF by the code key. That is often the method used to transmit CW signals; this one does an admirable job of it! In this fashion, the circuit merely has to produce a signal in the 700 to 1000 hertz audio frequency range, and that is sent out over the air. With this configuration, a single power transistor does the trick.

What we have here is a Pierce oscillator (see schematic diagram, Figure 10-1), whose tank circuit is comprised of coil T1 and capacitors C3 and C4. The crystal XTAL 1 in the 7000 to 7300 kilohertz region locks the transmitter onto its frequency and variable capacitor C2 trims the unit's output frequency.

Figure 10-1: 40-Meter Transmitter

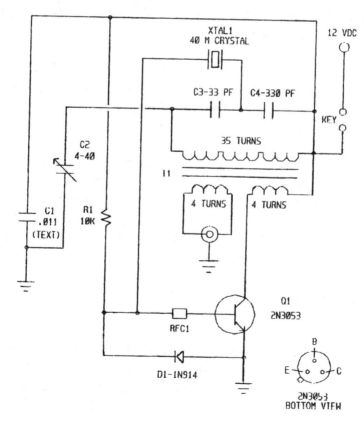

The first four-turn secondary of T1 is fed to the collector of transistor Q1, and this provides the needed feedback for the oscillator. The second four-turn T1 secondary provides inductive coupling for the antenna. The ferrite bead, RFC 1, is used for RFI suppression, capacitor C1 provides power supply bypass and resistor R1 furnishes circuit bias. And that is all there is to this device!

As stated above, the code key simply acts as an On-Off switch that allows the 12-volt direct current (DC) power source to reach the circuit. Hence, every time you hit the key, the signal will transmit for as long as the key makes electrical contact. By adjusting these transmission lengths, you can easily form *dits* (dots) and *dahs* (dashes) that will spell out your message.

I'll tell you fellow HAMers, *it donta get no simpler dan dis*. I can't help myself! Every so often, I just have to show off my command of the English language! Try to bear with me!

So, there are the basics behind this transmitting fool. Let me now talk about how to build it. There is only one even remotely difficult task — winding the coil — and even if you are a beginner, this will not give you any trouble.

Construction

In Figure 10-2, I have provided the printed circuit board (PCB) pattern. This can be used to make a PCB to hold the various components; however, this circuit can just as easily be wired point-to-point if desired. As I said, there is virtu-ally nothing that is critical about the transmitter.

Figure 10-2: 40-Meter Transmitter PCB Pattern

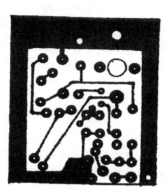

With your PCB in hand, install the capacitors, resistor, crystal socket, diode, and ferrite bead. A crystal socket is used to expedite changing frequency. Next, you will need to turn your attention to the coil. This is wound on a T50-2 toroid core (approximately ½ inch wide by ¼ inch thick, usually color-coded red) and consists of a 35-turn primary, wound evenly around the circular core, and two four-turn secondaries wound over the primary. The secondary windings need to be at roughly 60-degree angles from the primary winding's leads. Keep the four turns fairly close together, and before I forget, all windings use either number 24 or number 26 enameled coil (magnet) wire.

Try to be as accurate as possible in terms of getting the primary winding even and having the three at 60-degree angles, but this again is not terribly crucial. These frequencies tend to be fairly forgiving along these lines, but naturally, the better T1 is constructed, the better it will work.

Now that you have the coil constructed, clean (or CAREFULLY burn away) the enamel coating (insulation) on the leads and solder T1 in place. Do bear in mind the coil orientation when you install it. Next, it is time to solder transistor Q1 to the PCB. I have always left the more heat-sensitive and delicate semiconductor components for last. This is a personal preference that you may disregard, but I feel there is no need to expose these components to any more heat than necessary.

Again, observing the transistor's polarity, mount and solder the device about ¼ of an inch above the PCB. Figure 10-1 illustrates which lead is which on Q1. If you plan to operate at the 3-watt level, a small TO-5 heat sink would be a good idea for Q1. With that done, you have completed the parts placement of the printed circuit board. All that is left is to solder leads for the key 12 VDC power supply and antenna. Be sure to use 50 ohm coaxial cable when connecting the antenna to the T1 secondary.

If desired, the completed transmitter can be installed in a project case or other suitable enclosure. Power and phone jacks would do nicely for connection to the key and 12 VDC. I would recommend an SO-239 coaxial connector for the antenna. Also, if this suggestion strikes you, a milliamp meter could be hooked in series with the key. That would tell you how much current the unit is using when transmitting. And, if you want to get really fancy, a 0 to 15 DC voltmeter can be connected across the 12-volt supply to monitor the voltage. Additionally, you can dress up the front panel with a light emitting diode (LED) to indicate an ON-AIR condition.

Operation

I think we have pretty well covered this, but I will again touch on the highlights. Your key will be connected between the PCB and the positive side of a 12-volt direct current (DC) power source. The negative side will go to the board's ground rail. Next, hook an antenna to the transmitter. This can be anything from a simple dipole to a beam, as long as it is resonant on 40 meters.

Since the unit is crystal controlled (rock bound), it will be that frequency you will want to tune the receiver. Hit the key a couple of times and you should hear your signal. If all is well, you are ready to start calling CQ! Since you are restricted to a single frequency, that procedure will probably be the best way to make contacts. If at first you don't seem to attract any attention, don't despair! Keep trying, as I'm sure someone will soon hear you! Once you do get a reply, I think you will be surprised at the distance this transmitter will reach. Also, that first contact often spawns many more.

A close-up view of the very small 40-meter transmitter. Size and minimal part count make this an ideal unit for many HAM radio-related projects.

Conclusion

I have always had a special fascination with transmitters. I don't know, maybe it is because they are so simple to build. But, I rather suspect it is the ability to send voice, music, pictures, and other information out into thin air, and then be able to snatch it back and reproduce the original data. That, to me, borders on miraculous! So, getting into HAM radio has really helped promote a lifelong enchantment with the venerable transmitter.

Consequently, when I encounter a design like this one, I surely take notice. Actually, this is the composite of several designs that I have come across in the past. With a little manipulation of the capacitor values and the coil, it can be adapted to any or all the HF bands. That's what I mean about simplicity!

So, try this one out. It could come in handy at Field Day, during emergency operations, and just for fun! 40 meters lends itself well to this type of transmitter and this level of power, which further enhances the appeal of the unit!

Table 10-1

Tiny 40-Meter CW Transmitter Parts List

Semiconductors

Q1- 2N3053 NPN Power Transistor
D1- 1N914 Silicon Diode
XTAL1- 40 Meter Quartz Crystal (7,000 to 7,300 Kilohertz)

Resistors

R1- 10,000 Ohm ¼ watt Resistor

Capacitors

C1- 0.011 Microfarad Disk Capacitor (0.01 and 0.001 Disks Wired in Parallel)
C2- 4 to 40 Picofarad Compression Variable Capacitor
C3- 33 Picofarad Disk Capacitor
C4- 330 Picofarad Disk Capacitor

Other Components

T1- 35-Turn Primary and two Four-Turn Secondaries Wound on a T50-2 Toroid Core with #24 or #26 Enameled Coil Wire (See Text)
RFC1- Small Ferrite Bead
ANT- SO-239 Female Coaxial Connector

Conclusion

I have been involved in HAM radio for a little while now and have no regrets about my decision to get involved. It is, to me, an engaging and fascinating hobby that offers something for everybody. That is, if you are interested in radio. But, I'm certain you are!

I have had tremendous fun writing this book and hope you have had at least that much fun reading it! It provided me an opportunity to meet a number of very interesting people and learn an extraordinary amount about both the hobby, and electronics in general. For that alone, this text has been well worth the effort it entailed.

However, I also truly hope it has conveyed to you some of the same affection and respect I have gained for amateur radio. After all, that is the single most important reason for this guide. In the process of compiling the book, I found a whole new world out there called HAM radio. It is a world I did not know existed. It is my desire that you also discover that world!

Over the years I have been immersed in electronics, I kept toying with the idea of becoming a HAM. For years, the fear of Morse code kept me away. Finally, when the no-code Technician ticket came along, I no longer had an excuse not to join the hobby. Actually, I was very enthusiastic about the prospect of finally achieving a long desired goal. What I found in amateur radio, when I got there, was anything but what I expected. The traditional attitude HAMs have toward both the hobby and each other was refreshing and is something I have grown to admire and depend on.

Hence, I eagerly await your presence in our ranks. Aside from the vast number of activities amateur radio affords, the prevailing camaraderie, co-operation and widespread goodwill is nothing short of impressive! I know it impressed me, and I'm hard to impress! Well, so some people say! Hum....what do they know?! Right? Right!

Oh well, enough of that! Try it for yourself! It is easier than ever to become a HAM, and once you have your license, you will certainly be welcomed with open arms. The world of amateur radio can be a tightly knit one, but newcomers are the hobby's life blood. It has been my experience that virtually all HAMs recognize that fact.

To keep our beloved hobby dynamic, new individuals are both desired and necessary. Trust me on this, even the old timers are quick to appreciate the contribution of folks just getting into amateur radio. Not only do they appreciate it, they salute a rookie's viewpoint as a chance to expand on their own perspective of the hobby. There is one thing you find with most HAMs; they are people willing and eager to learn!

On that note, let me again encourage you to join one of the most enchanting enterprises around. At least, that is how I find HAM radio to be. I believe you will discover it is everything I have praised it for and much more. Really, don't you owe yourself the opportunity to find out? Sure you do! So then I will see you on the air!

Carl KG4AIC

Source List

Here is a list of some of the many manufacturers, dealers and other outlets for amateur radio equipment. Although I have not as yet dealt with everyone on this list, all advertise in major magazines, so I expect they are reliable sources. Hope they are of help to you in your quest for HAM gear!

ADI Corp./Pryme Communciations Corp.
480 Apollo Street #E
Brea, CA 92821
Orders: (800) 666-2654
e-mail: premier@adi-radio.com,
Internet: www.adi-radio.com
*Amateur radio manufacturer

Advanced Battery Systems, Inc.
300 Centre Street
Holbrook, MA 02343
Phone: (800) 634-8132
e-mail: periphex@aol.com
Internet: www.advanced-battery.com
*Batteries

Advanced Receiver Research
Box 1242
Burlington, CT 06013
Phone: (860) 485-0310
*VHF/UHF/Microwave equipment

Amateur Electronics Supply (AES)
5710 West Good Hope Road
Milwaukee, WI 53223
Order: (800) 558-0411
e-mail: help@aesham.com
Internet: www.aesham.com
*Amateur radio equipment - new, used

Alinco Electronics, Inc.
438 Amapola Avenue, Suite 130
Torrance, CA 90501
Phone: (310) 618-8616
Internet: www.alinco.com
*Amateur radio manufacturer

All Electronics, Inc.
Post Office Box 567
Van Nuys, CA 91408-0567
Order: (800) 826-5432
e-mail: allcorp@allcorp.com
Internet: www.allcorp.com
*Electronic parts and surplus equipment

Alpha Delta Communications, Inc.
Post Office Box 620
Manchester, KY 40932
Orders: (888) 302-8777
Internet: www.alphadeltacom.com
*Antennas

Alpha/Power, Inc.
14440 Mead Court
Longmont, CO 80504
Phone: (970) 535 4173
Internet: www.alpha-power-inc.com
*Linear amplifiers

Ameritron
116 Willow Road
Starkville, MS 39759
Free Catalog: (800) 713-3550
Internet: www.ameritron.com
*Linear amplifiers and meters

American Radio Relay League (ARRL)
225 Main Street
Newington, CT 06111-1494
Phone: (860) 594-0355
Orders: (888) 277-5289
e-mail: pubsales@arrl.org
Internet: www.arrl.org
*Primary benefactors and lobby group for amateur radio. Texts and study materials. QST magazine.

Antique Radio Classified
Post Office Box 802-B22
Carlisle, MA 01741
Phone: (978) 371-0512
*Magazine about antique radio. Free sample copy.

Associated Radio Communications
8012 Conser
Overland Park, KS 66204
Phone: (800) 497-1457
Internet: www.associatedradio.com
*Amateur radio equipment - new and used

Austin Amateur Radio Supply
5325 North I-25
Austin, TX 78723
Phone: (800) 423-2604
Internet: www.aaradio.com
*Amateur radio equipment - new only

Burghardt Amateur Supply, Inc.
710 10th Street SW
Watertown, SD 57201
Phone: (605) 886-7314
Orders: (800) 927-4261
e-mail: hamsales@burghardt-amateur.com
Internet: www.burghardt-amateur.com
*Amateur radio equipment - new and used

Cable X-Perts, Inc.
416 Diens Drive
Wheeling, IL 60090
Phone: (847) 520-3003
Orders: (800) 828-3340
Internet: www.cablexperts.com
*Coax and other feed line

Command Technologies, Inc.
1207 West High Street
Bryan, OH 43506
Phone: (800) 736-0443
e-mail: cmdrtech@bright.net
Internet: www.command1.com
*HF and VHF Linear amplifiers

Communications Electronics, Inc.
Post Office Box 1045
Ann Arbor, MI 48106-1045
Phone: (734) 996-8888
Orders: (800) USA-SCAN (872-7226)
e-mail: cei@usascan.com
Internet: www.usascan.com
*Scanners and other radios

Digi-Key Corporation
701 Brooks Avenue South
Thief River Falls, MN 56701-0677
Phone: (800) 344-4539
Internet: www.digikey.com
*Electronic components

Gap Antenna Products, Inc.
99 North Willow Street
Fellsmere, FL 32948
Phone: (561) 571-9922
Internet: www.gapantenna.com
*HF and VHF antennas.

The HAM Contact
Post Office Box 4025
Westminser, CA 92684
Phone: (714) 901-0573
Order: (800) 933-4264
Internet: www.hamcontact.com
*Portable power sources

HAM Radio Outlet (HRO)
933 North Euclid Street
Anaheim, CA 92801 (there are 12
stores nation wide)
Phone: (714) 533-7373
Orders: (800) 854-6046
Internet: www.hamradio.com
* Amateur radio equipment

The HAM Station
220 North Fulton Avenue
Evansville, IN 47719-0522
Phone: (800) 729-4373
Internet: www.hamstation.com
*Amateur radio equipment - new
and used

HAMtronics, Inc.
65 Moul Road
Hilton, NY 14468-9535
Phone: (716) 392-9430
e-mail: jv@hamtronics.com
Internet: www.hamtronics.com
*VHF/UHF radio equipment

hy-gain
300 Industrial Park Road
Starkville, MS 39759
Phone: (800) 647-1800
e-mail: www@hy-gain.com
Internet: www.hy-gain.com
*Antennas for all bands. Free catalog.

ICOM America, Inc.
2380 11th Avenue
Bellevue, WA 98004
Phone: (425) 454-8155
Internet: www.icomamerica.com
*Major amateur radio manufacturer

Jameco Electronics, Inc.
1355 Shoreway Road
Belmont, CA 94002
Phone: (800) 831-4242
Internet: www.jameco.com
*Electronics components, kits,
computer gear and test equipment

JDR Microdevices
1850 South 10th Street
San Jose, CA 95112-4108
Phone: (800) 538-5000
Internet: www.jdr.com
*Electronic parts, kit, equipment and
computer gear

JUN'S
5563 Sepulveda Blvd.
Culver City, CA 90230
Phone: (310) 390-8003
e-mail: radioinfo@juns.com
Internet: www.juns.com
*Amateur radio equipment - new
and used

Kenwood Communications Corporation, Amateur Radio Products Group
2201 East Dominguez Street
Long Beach, CA 90801-5745
Phone: (310) 639-5300
Internet: www.kenwood.net
*Major amateur radio manufacturer

Lentini Communications, Inc.
21 Garfield Street
Newington, CT 06111
Phone: (800) 666-0908
Internet: www.lentinicomm.com
*Amateur radio equipment - new only

Maha Communications, Inc.
2841-B Saturn Street
Brea, CA 92821
Phone: (800) 376-9992
Internet: www.maha-comm.com
*Batteries and battery chargers.

MFJ Enterprises, Inc.
Post Office Box 494
Mississippi State, MS 39762
Phone: (601) 323-5869
Internet: www.mfjenterprises.com
*Manufacturer and distributor of amateur radio equipment

Mirage Communications Equipment
300 Industrial Park Road
Starkville, MS 39759
Phone: (601) 323-8287
e-mail: www@mirageamp.com
Internet: www.mirageamp.com
*Mobile and base station linear amplifiers

Mouser Electronics
958 North Main
Mansfield, TX 76063-4827
Phone: (800) 346-6873
Internet: www.mouser.com
*Electronic components and equip.

P.C. Electronics
2522 South Paxson Lane
Arcadia, CA 91007
Phone: (626) 447-4565
e-mail: tom@hamtv.com
Internet: www.hamtv.com
*Amateur television equipment

R & L Electronics
1315 Maple Avenue
Hamilton, OH 45011
Orders: (800) 221-7735
e-mail: sales@randl.com
Internet: www.randl.com
*Amateur radio equipment

Radio City, Inc.
2663 County Road I
Mounds View, MN 55112
Phone: (800) 426-2691
e-mail: radiocty@skypoint.com
Internet: www.radioinc.com
*Amateur radio equipment

Radio Shack (Tandy Corporation)
Local Stores nationwide or Fort Worth, TX 76102
*Parts, hardware, cases and some amateur radio equipment

RF Parts
435 South Pacific Street
San Marcos, CA 92069
Phone: (760) 744-0700
e-mail: rfp@rfparts.com
Internet: www.rfparts.com
*Radio frequency tubes, transistors, power modules

RT Systems
8207 Stephanie Drive
Huntsville, AL 35802
Phone: (800) 723-6922
e-mail: sales@rtsars.com
Internet: www.rtsars.com
*Amateur radio equipment

Supercircuits
One Supercircuits Plaza
Leander, TX 78641
Orders: (800) 335-9777
Internet: www.supercircuits.com
*B&W and color television cameras,
transmitters, down converters for
ATV applications

Ten-Tec, Inc.
1185 Dolly Parton Parkway
Sevierville, TN 37862
Orders: (800) 833-7373
e-mail: sales@tentec.com
Internet: www.tentec.com
*Amateur radio kits and equipment

Texas Towers (RF Distributors, Inc.
1108 Summit Avenue, Suite #4
Plano, TX 75074
e-mail: sales@texastowers.com
Internet: www.texastowers.com
*Antenna towers and amateur radio
equipment

The Radio Works
Post Office Box 6159
Portsmouth, VA 23703
Orders: (800) 280-8327
e-mail: jim@RadioWorks.com
Internet: www.radioworks.com
*Antennas and antenna supplies

**Tornonto Surplus & Scientific,
Inc.**
608 Gordon Baker Road
Willowdale, Ontario
Canada M2H-3B4
Phone: (416) 490-8865 (store)
(905) 887-0007 (office)
Internet: www.torontosurplus.com
*Surplus test equipment, radio gear
and other electronics

Universal Radio, Inc.
6830 Americana Parkway
Reynoldsburg, OH 43068
Orders: (800) 431-3939
e-mail: dx@universal-radio.com
Internet: www.universal-radio.com
*Amateur radio equipment

Vectronics
1007 Highway 25 South
Starkville, MS 39759
Phone: (601) 323-5800
Orders: (800) 363-2922
Internet: www.vectronics.com
*Amateur radio kits, antenna, tuners
and meters

W & W Manufacturing Company
800 South Broadway
Hicksville, NY 11801-5017
Phone: (800) 221-0732
e-mail: w-wassoc@ix.netcom.com
Internet: www.wwassociates.com
*Batteries, battery packs and holsters

Yaesu USA
17210 Edwards Road
Cerritos, CA 90703
Phone: (562) 404-2700
Internet: www.yaesu.com
*Major amateur radio manufacturer

---

*This is only a partial list of the manu-
facturers and distributors that serve
the amateur radio community. If I have
missed your favorite, I apologize!
Many of these companies supply either catalogs or other information concerning their products upon request.
Check with the ones you're interested
in via telephone or the Internet.*

Study Materials

American Radio Relay League (ARRL)
225 Main Street
Newington, CT 06111-1494
e-mail: pubsales@arrl.org
Internet: www.arrl.org

The W5YI Group, Inc.
P.O. Box 565101
Dallas, TX 75356
Phone: 800-669-9594

Radio Shack
Local Stores

Code Quick from Wheeler Applied Research Lab
38221 Desert Greens Drive West
Palm Desert, CA 92260-1005
Phone: 619-773-9426 or 619-773-5795

HAM Radio Outlet (HRO)
933 North Euclid Street
Anaheim, CA 92801
Phone: Western 800-854-6046
       Mountain/Central 800-444-9476
       South East 800-444-7927
       Mid-Atlantic 800-444-4799
       North East 800-644-4476
       New England/Canada 800-444-0047
Internet: www.hamradio.com

Amateur Electronic Supply (AES)
5710 West Good Hope Road
Milwaukee, WI 53223
Phone: 800-558-0411
e-mail: help@aesham.com
Internet: www.aesham.com

There are many other sources for study guides, informational materials, and code study courses. These are a few that are know to have excellent reputations for providing materials that have successful results. If possible, try the program before you buy it to make sure it will work for you.

Internet
Address Book

This is a list of interesting amateur radio-related Internet sites. Here again, I have listed the ones I have used or that have been recommended to me, but there are many, many more. If you have a favorite that is not listed, please send it to me c/o Howard W. Sams, Prompt Books division.

In each case, the address is given as it should be entered on your keyboard. Please note that one address does have capital letters. The information after the hyphen is not part of the address, but only a brief description of what is found on the site. Do not enter this when contacting a site.

www.adi-radio.com - ADI/PRYME® Amateur Radio Equipment Homepage

www.alinco.com - Alinco® Electronics Homepage

www.amradiotrader.com - Amateur Radio Trader Magazine Homepage

www.amsat.com - Satellite Information

www.arrl.org - ARRL (American Radio Relay League) Homepage

www.artscipub.com - Books & QSL cards

www.batteriesamerica - Batteries and Accessories

http://central.alabama.skywarn.net - Sample Skywarn Page

www.dransom.com - STS+ Satellite Tracking (download latest version)

www.ebay.com - Auction Site, Handles HAM Equipment

www.fcc.gov - FCC Homepage

www.hammall.com - Equipment Sales

www.icomamerica.com - ICOM Homepage

www.kenwood.net - Kenwood Homepage

http://liftoff.msfc.nasa.com/REALTIME/JTRACK/ amateur.html - Satellite Tracking

www.logsat.com - Satellite Tracking

www.nasa.gov - NASA Homepage

www.packetradio.com - Packet Radio Homepage

www.qrz.com - HAM Information and Call Sign Directory

www.siliconpixel.com - W95SSTV Slow Scan TV Site

www.terraserver.com - Maps

http://web.usna.navy.mil/~bruninga/aprs.html - APRS site

www.yaesu.com - Yaesu Amateur Radio Equipment

Acknowledgments

I would like to thank the following for assistance in producing this text. Their help was eagerly sought out and gratefully appreciated.

AMERICAN RADIO RELAY LEAGUE

COMPTON'S INTERACTIVE ENCYCLOPEDIA

WEBSTER'S NEW WORLD DICTIONARY, THIRD COLLEGE EDITION

HAM RADIO OUTLET

THE MONTGOMERY AMATEUR RADIO CLUB

ALL ABOUT HAM RADIO. By Harry Helms AA6FW. High Text Publications, 1992.

TECHNICIAN NO-CODE PLUS. By Gordon West WB6NOA. Master Publishing, 1997.

KENWOOD USA

I would also like to thank all the HAMs in my local area for their generous contributions to this effort and the following for service above and beyond the call of duty. Without the help of everyone involved, this text would not have been possible.

Fred Beatty K8AJX   *Jim Rice KR4WN*

Don Dahl W3MK   *Mike Meriwether KT4XL*

Pete Carroll KF4QOE   *Dennis Rumbley KS4UO*

Bobby Chandler N4AU

Again, my sincerest thanks to all who assisted!

Glossary

This is a list of common terms as they apply to amateur radio. In many cases, the terms have the same meaning for both HAM radio and electronics, but there are some that take on an altogether different connotation when pertaining to amateur radio. This list should provide the definitions needed to better understand the hobby.

A

ADVANCED - This is a now-defunct license class, but any HAM who held the Advanced class prior to April 15, 2000, can retain that class indefinitely.

AMATEUR RADIO SERVICE - The official name for HAM radio. The title symbolizes a demeanor of community service shared by most amateur radio operators.

AMATEUR PACKET REPORTING SERVICE (APRS) - A packet based positioning system used by amateur radio. Allows base and mobile stations to be precisely situated on computer generated maps.

AMATEUR TELEVISION (ATV) - A system of transmitting live television pictures over amateur radio frequencies. Like commercial television, each channel uses a 6 megahertz spectrum.

AMPERE - Named after the French physicist A. M. Ampere, it is the basic unit of current. Often expressed in milliamperes, it is important in power calculations.

AMPLIFIER - An electronic device that increases a signal's strength. The signal can be audio or radio frequencies.

AMPLITUDE MODULATION (AM) - A form of modulation of a carrier in which the height (amplitude) of the waveform (cycles) changes. Seen mostly in high-frequency (HF) radio.

ANALOG - Having to do with a broad spectrum of values as opposed to the two units (0 and 1) of digital. Most amateur radios use an analog scheme in their operation.

ANTENNA - The device that is used to radiate a transmitted signal, or receive a signal radiated from another station. These come in a multitude of sizes and designs, but are an important part of the efficiency of all radio stations.

AUDIO - The frequencies the human ear is able to hear (usually about 20 to 20,000 hertz). Also, the sound produced by your radio.

B

BALUN - A device normally made up of coils and/or capacitors that balances (matches) the impedance of an antenna system to the feed line being employed. For example, matching a 300-ohm antenna to a 50-ohm radio.

BAND - Any one of numerous frequency allocations as set up by the Federal Communications Commission (FCC). Some examples are the 40-meter, 10-meter and 2-meter bands.

BEAM - A type of antenna that sends most of the transmitted signal in one direction. It also receives said signal from one direction. A very popular concept to obtain greater range or locate a station of unknown position.

BOOM - This is a loud noise made when something explodes. True, however in our context, this is the center support of a quad or Yagi beam antenna.

BRAID - This is the woven outer shield element of coaxial cable. It is usually used for the ground connection.

C

CABLE - In HAM radio, this refers to the feed line or lead-in wire connecting the radio to the antenna. Can be coaxial, ladder line, etc.

CALL SIGN - The number/letter combination issued by the FCC to a HAM for identification. For example, mine is KG4AIC. Vanity call signs can be requested from the commission, but there is a fee, and only available calls will be issued. To many HAMs, this is as important as their name. It is also important that you give your call sign at least every 10 minutes.

CAPACITOR - An electronic component that stores an electrical charge and also blocks direct current.

CARRIER - This is the backbone of most transmitted signals. That is, the modulation (voice, CW, picture) rides piggyback on the carrier signal. One exception to this is single sideband (SSB) where the carrier and one side of the cycle is deleted.

CLASS - In amateur radio, this refers to the degree of your license. At present, there are three classes of license; Technician, General, and Extra.

COAXIAL - Often referred to as just coax, this is a shielded style antenna feed line. The shield is normally grounded, and coax comes in either 50- or 75-ohm impedances. It is also one of the most popular antenna cables.

CODE - This is short for Morse Code, and is the universal language of "dits" and "dahs" used by many HAMs to convey information.

COIL - Also called an inductor, coils are successive windings of wire that generate a magnetic field. These are important to electronics in a number of ways, including frequency tuning and relay activation.

CONNECTOR - A device that joins two or more pieces of electronic gear together. These can be antenna, microphone, BNC, PL-259 and so forth, connectors. They also are designated as male and female depending on their design.

CONTEST - Yah Boy! That's where you win the blue ribbon at the fair! While that might be, contests to HAMs are competitions in which operator skills are demonstrated. This often involves making as many contacts with other stations as possible in a given time period.

CONTINUOUS WAVE - Better know as CW, this is the oldest and still a very popular form of amateur radio communications. It involved conveying information using Morse code.

CRYSTAL - A frequency determining component made from quartz. The quartz is thinly sliced and connected to electrodes, and when electricity is applied, the quartz vibrates at a specific rate. That rate is the frequency of the crystal.

CUBICAL QUAD - A type of directional, or beam antenna that consists of two or more boxes made from wire and supports. When assembled, the antenna takes on a cubical appearance.

D

DIGITAL - Dealing with values of only "0" (low) and "1" (high). In the digital scheme, everything is forced to one of those two values, unlike analog where there can be numerous values between 0 and 1. Digital is the coming age and is being seen more and more in amateur radio.

DIGITAL SIGNAL PROCESSING (DSP) - This is an electronic method of processing signals to improve intelligibility. By eliminating noise and other distractions, a signal can often be better understood.

DIODE - An electronic component that allows the flow of electricity in only one direction. Used for rectification and frequency tuning.

DIPOLE - An antenna design that utilizes two sections of wire connected to the feed line in the middle. Each wire section is cut to the wavelength of the frequency being transmitted or received, and one side acts as the radiator while the other is the radial. A very efficient and simple system loved by many HAMs.

DIRECTOR - This is the element(s) of a beam antenna that aim the signal. In a simple Yagi, this will be the element after the driven element. In more complex systems, the antenna may have many director elements.

DRIVEN ELEMENT - This is the element of a beam antenna that actually radiates the signal. Usually made up of an individual radiator and individual radial, with UHF it can be a single piece of wire.

DUMMY LOAD - In HAM radio, this is a substitute antenna used for tuning purposes. A dummy load absorbs most of the transmitter's signal without causing harm to the radio. However, it will radiate a slight signal.

E

EARTH-MOON-EARTH - An area of HAM radio where the operator bounces the signal off the moon and back to a distant part of Earth. Also known as "Moon Bounce".

ELEMENT - In this context, an element is one of the active components of an antenna system. Example; driven element of a beam.

EXTRA - In HAM radio, this is the highest license class that can be obtained and carries the maximum number of privileges.

F

FEDERAL COMMUNICATIONS COMMISSION (FCC) - Ooooh!!! This is the "Big Kahuna"! It is, in fact, the governing body of all radio frequency transmissions, and sets the rules that will be followed. In all fairness, the FCC does look out for the interests of radio operators and without the commission, the "air waves" would be in total kayos.

FIELD STRENGTH METER - An electronic device that measures the strength of a nearby radio transmitter's signal. That is, the field strength of the signal.

FILTER - In this context, a filter is an electronic circuit that restricts certain frequencies from being transmitted or received. Two common radio filters are low-pass and bandpass.

FREQUENCY - Expressed in hertz (kilohertz, megahertz, gigahertz) this is the wavelength of a transmitted signal. For example, 147.555 megahertz is a common frequency in the 2-meter band.

FREQUENCY MODULATION (FM) - This is a form of modifying a carrier wave to convey information in which the height (amplitude) of the cycles remains the same, but the frequency of those crests varies according to the modulation signal. This is seen more often in the VHF and UHF bands.

G

GEAR - In HAM lingo, this is your stuff. OK, this represents all the equipment you have and use in your station, or shack.

GENERAL - While this a military rank of high esteem, to a HAM this is the second level in the license class structure. That is, as of this writing. The general class license provides the largest jump in operating privileges of the three classes.

GROUND - This is the place all electricity wants to go. Sort of like Disney World. I'm not kidding! Electricity is trying to find the path of least resistance to ground. In HAM radio, this can literally mean the earth. Antennas and radio systems are usually connected to the earth through grounding rods driven into the soil. It is also seen as a common bus in electronic circuits (usually the negative side), which, by the way, sometimes ends up connecting to the grounding rod.

GROUND PLANE - A type of omnidirectional antenna that employs a single vertical radiator, and three or four radials set at an angle to the sides of the radiator. This is an excellent design and quite popular with HAMs.

H

HAM - A nickname, that carries great respect, for amateur radio operators. Its origin is somewhat cloudy, but one explanation says this was the call sign of one of the very first operators, and just caught on. Who knows for sure?

HAND HELD TRANSCEIVER (HT) - A small transceiver that is capable of limited range communications. They are easy to carry around, and their range will vary with the frequency they operate on, and the antenna they use. A big favorite among most HAMs.

HERTZ - Named after German physicist Heinrich Hertz, this is the unit of frequency. For many years, the hertz was known as the cycle.

HIGH FREQUENCY (HF) -This is collectively the lower frequencies used by the amateur radio service. Generally speaking, these are from 1.8 MHz to around 30 MHz (160 to 10 meters).

I

INDUCTOR - AKA (also known as) the coil, this is a series of sequential wire windings that generate a magnetic field. Used in frequency control, antenna matching and filter systems, just to name a few.

INTEGRATED CIRCUIT (IC) - These are marvelous little electronic components, often called chips, that are an almost entire circuit on a microscopic piece of silicon wafer. They have revolutionized electronic since their introduction around the early 1960's.

K

KEY - The name given to the device that taps out Morse code. They come in straight keys and as paddle designs and are the staple of HAMs who enjoy code.

L

LADDER LINE - A type of feed line that looks like a ladder. The two conductors are on the outside (legs of the ladder) attached together with pieces of plastic (rungs of the ladder).

LASER - Acronym for Light Amplification by Stimulate Emission of Radiation, in HAM radio this device is used for light communications. The laser is capable of a very tight (coherent) beam of light, making it ideal for this purpose.

LICENSE - Also known as your ticket, this is the piece of paper, issued by the FCC, that allows you to operate your transmitters. Of course, you earn the license through tests, and they presently come in three flavors.

LIGHT EMITTING DIODE (LED) - A semiconductor device that emits light when voltage and current are passed through it. They are normally comprised of a Gallium-Arsenide junction that posseses the light emission property.

LINEAR AMPLIFIER - These are power amplifiers that boost your transmitter's output. Handy when greater range is needed, the linears come in various sizes and some can increase the power to the legal limit of 1500 watts.

LONG WIRE - This is a name for an antenna that consists of long pieces of wire. The dipole is a form of this system. These are often used for receiving purposes, as with receiver, the wavelength isn't as important, so the antenna doesn't have to be a specific length.

M

METER - This is a device that is used to measure a given quantity. They can be mechanical (analog) in nature or electronic (digital). They can measure, amperage, voltage, wattage, SWR and the like.

MICROPHONE - An electronic device that detects sound and turns it into electrical impulses. In HAM radio, this is the device that allows for "phone" transmissions.

MICROWAVES - I will probably get an argument here (I usually do), but any frequency over the 1-gigahertz mark is, to me, in the microwave range. This region extends all the way to light.

MOBILE - In HAM lingo, this is a nonfixed station. That is, a station that has the ability to change locations. You know, like your car. Most mobile stations are in a vehicle of some sort.

MOON BOUNCE - Also known as earth-moon-earth, this is bouncing your signal off the moon and back to earth. Fantastic ranges can be reached, but it does take power.

MORSE CODE - Developed by Samuel Morse, this is the "dit/dah" method of sending information over telephone wires, or in our case, radio. Morse code has long been one of the test elements to obtain certain HAM licenses.

N

NOVICE - A now-defunct license class, this was the entry level for amateur radio. However, any HAM that held this class prior to the April 15, 2000, restructuring can retain the ticket as long as he or she desires.

P

PACKET - A digital form of information transfer, this is the method HAMs use to send information from their computers out over the air. Using a device called a terminal node controller (TNC), their radio becomes a wireless Internet of sorts.

PADDLE - A form of key for Morse code that allows the operator to squeeze two paddles together rather than pushing a lever down. These come with names like Vibrokeyer, Lambic, Brass Racer and, of course, Original. They are considered by many HAMs to be superior to the original Straight Key.

PHONE - In our context, this means voice transmissions. The second oldest form of HAM communications, it is probably the most popular today.

POTENTIOMETER - A variable resistor. Used for volume, squelch and tuning controls, among others, this is a very handy electronic component.

POWER SUPPLY - This is the source of electric power for your amateur equipment. Some gear has them built in, while other equipment requires an external supply. But, one thing is for sure! Your gear isn't going to work at all without a power supply.

POWER - In this context, this is the output level of a transmitted signal. Designated in watts, or a version thereof (milliwatts), this tells you the amount of power the transmitter delivers. It can also apply to energy consumption, as in a 100-watt lamp.

PRIVILEGE - Each class of amateur radio license comes with privileges. These are the bands you can work, and what type of operation is permitted (phone, CW, etc.). Naturally, the higher the class, the more privileges.

PROPAGATION - To a HAM, this is the atmospheric conditions that allow their signals to either travel far, or not so far (restricts them). This involves such things as sun spots, skip and iono-spheric ionization, all of which have an effect on the distance your signal will go.

Q

QRP - This is the "Q" code designation for low power operation. Usually, QRP stations are 5-watts of power or less.

QSL - This is the "Q" code designation for acknowledgment of contact. Also applies during QSOs as "Do you understand?" QSL is additionally used to designate written confirmation in the form of a QSL card sent by one HAM to another following the con-tact.

QSO - Pronounced "Q-So", this is the designation for a contact and/or conversation between two stations. This can be either by phone (voice) or by CW (Morse code), as well as other formats.

R

RADIAL - This is the grounded side of the antenna that reflects the signal as opposed to radiating it. That is, it is usually grounded, but there are a few other designs.

RADIATOR - This is the "hot" side of most antennas that actu-ally emits the transmitter signal. They also receive the incoming signals in conjunction with the radial(s).

RADIO - In HAM terms, this is anything that receives and/or transmits a signal. Among others, that includes receivers, trans-mitters and transceivers.

RADIO FREQUENCY (RF) - The spectrum of electromagnetic waves that covers radio transmissions. This is normally consid-ered to be from around 50 kilohertz to over 100 gigahertz, give or take a few megahertz here and there.

RADIO FREQUENCY RELAY - An electronic device that "sniffs out" radio frequency energy (RF) and activates a relay when such energy is present. The relay can then control other devices.

RADIO TELETYPE (RTTY) - Pronounced "ritty", this is a method for sending data by radio in a coded format. RTTY has been used for many years and is still a favorite among a lot HAMs. It is a fast and accurate way to convey said information.

RECEIVER - Any electronic device that intercepts, detects and interprets a transmitted radio signal. These come in all types and shapes, from basic receivers to scanners, computer-controlled devices and wideband radios.

REFLECTOR - In our context, this is the element of a beam antenna that reflects the signal. Usually a single element next to the radiator/radial combination (driven element).

RELAY - A magnetically oriented device that acts as a switch when activated. The electromagnet causes switch contact to move (close) when energy is passed through its coil.

REPEATER - In HAM terminology, this is an automatic station that receives a signal on one frequency, then sends it out (repeats it) on another frequency. Used heavily in VHF and UHF, these stations increase the range and clarity of incoming signals.

RESISTOR - An electronic component that resists the flow of electricity. These are used for bias (voltage) control, and sometimes as part of a timing circuit.

RIG - HAM lingo for your radio equipment. Transmitters, receivers and transceivers all fit this category.

ROTOR - An electromechanical device that aims an antenna. Electricity is sent to a motor that turns the antenna, thus pointing it in the direction desired by the operator. Almost essential for beam-style antennas.

S

SATELLITE - Any body orbiting the Earth. For HAMs, however, this has a more substantial meaning. Many satellites are equipped with amateur radio gear and can be used as flying repeaters. This part of the hobby has a big following.

SCANNER - A specialized receiver that sequentially checks a large number of frequencies for activity. These come in varying degrees of complexity, but all "scan" frequencies programmed into them.

SHACK - This is where a HAM keeps his or her equipment. It may be the most elegant part of the home, but is still called the HAM Shack!

SIGNAL - Any electromagnetic energy coming into a receiver, or being sent by a transmitter. This can be just a carrier or modulated to convey information.

SINGLE SIDEBAND (SSB) - A form of radio transmission in which the carrier and one half of the waveform have been suppressed (removed) from the transmitted signal. Very popular format for HF and other areas of the amateur radio spectrum.

SLOW-SCAN TELEVISION (SSTV) - A method of sending still pictures by radio. This has been around for quite some time, and due to the small bandwidth required, is seen on many HAM bands.

SPEAKER - Not that boring guy at the club luncheons! This is an electromagnetic device that turns electrical impulses into audible sound. One of these is found on most receivers.

STANDING WAVE RATIO (SWR) - This is the ratio between the amount of a transmitted signal that leaves the antenna and goes into the air, and the amount that remains (stands) on the feed line. These are designated as "#:#", where 1:1 is great (flat) and 5:1 is BAD! So bad that it can and will do damage to your transmitter.

T

TECHNICIAN - In the HAM licensing ladder, this is the first rung. It requires passing a written exam, but no code test. A second version of this class is the Technician with HF privileges. That comes when a Technician has also passed the Morse code test.

TECHNICIAN PLUS - A now-defunct license class as of the April 15, 2000, license restructring. Unlike the Novice and Advanced classes, this one is kaput!

TERMINAL NODE CONTROLLER (TNC) - This is the electronic device that goes between a computer (PC) and your radio when you want to transmit the information on the computer. It is basically a specialized modem that changes the computer's digital signals to analog so they can be transmitted by radio.

TICKET - In HAM lingo, this is your license to operate. I haven't been able to find the origin of this term, but that's what it is.

TOWER - A reinforced structure that extends skyward and holds your antenna(s). Naturally, it holds them up high. That's the idea! These come in sizes from 20 to 30 feet high to over 100 feet, and their construction allows you to climb them. Also, some will crank up and down while others lean over for better access to the antenna(s).

TRANSCEIVER - A combination transmitter and receiver all in one package. These employ electronic means to change from one mode to the next, and due to convenience and economic considerations, have become very popular.

TRANSISTOR - A semiconductor device that can be employed as a switch or an amplifier. Small in size, dependable and cool in operation, they have all but replaced vacuum tubes except in high-power applications.

TRANSMITTER - The electronic device that send your signal out over the air. This is accomplished by converting the information you want to send into electromagnetic waves of a specific frequency. Individual transmitters are not as common as they once were due to the popularity of transceivers.

TUNER - In HAM radio, this usually refers to an antenna tuner. This is a device that matches the impedance of an antenna to a radio, and also allows adjustment of the SWR.

U

ULTRAHIGH FREQUENCY (UHF) - I consider this the spectrum of frequencies from about 300-megahertz to 1-gigahertz. I do, however, get an argument on that from time to time. Anyway, it is in that general range.

V

VERY HIGH FREQUENCY (VHF) - Here again, I get arguments on this, but I consider VHF to be the frequencies between 30-megahertz and 300-megahertz. This is actually broken up into two bands: VHF and LVHF (Lower VHF).

VERY LOW FREQUENCY (VLF) - This is a seldom used band that is well below the AM broadcast band. Usually this is the 160-to 190-kilohertz range and is plagued with noise.

VOICE - Another name for phone, this is the process of sending information by your voice as opposed to CW.

VOLT - Named after the Italian physicist Count Alessandro Volta, this is the unit of electromotive force (potential).

VOLUNTEER EXAMINER (VE) - This is the person that volunteers to give FCC examinations. We all owe these folks a great debt of gratitude as they have made getting our licenses far more convenient and much quicker than the old FCC system. Believe me, you don't even want to hear about the old system!

W

WATT - Named after Scottish scientist James Watt, this is a unit of power. Defined as amperage times voltage, it specifies the brute force behind the voltage.

WAVELENGTH - In HAM radio, this usually refers to the band on which you are operating. However, it can also refer to the frequency of operation. Scientifically, it is defined as the distance between two crests of a waveform.

There you have it. The terms you hear and use in your pursuit of amateur radio. Learn them, use them but most of all, have fun!

Index

Symbols

U

UHF 67, 79, 104, 252,
  256, 264, 379
Ultrahigh Frequency 67, 379
unity gain 97, 98, 293
upper sideband 63, 232
USB 63

V

vacuum tube volt meter 128
VE 20, 380
VEC 20
Vectronics 327
Very High Frequency 66, 380
Very Low Frequency 69, 380
VHF
  66, 97, 104, 251, 256, 380
VLF 69, 70, 380
voice communications 61

voltmeter 128
Volunteer Examiner 20, 380
Volunteer Examiner
  Coordinator 20

W

wallpaper 230
WARC 262
watt meter 113, 128
wavelength 236, 243, 380
weak-signal communications 71
whip 99
World Authority Radio
  Conference 262

Y

Yagi 46, 93, 238
Yagi beam 242
yarg 234

Sourcebook for Electronics Calculations, Formulas, and Tables

by Newton Braga

This book is written for the engineer, student, technician, or hobbyist who designs or needs to understand more about electronic circuits. Newton Braga has compiled an assortment of all the basic information necessary to make calculations when designing new projects. Instead of searching through a maze of old school or reference books, turn to the *Sourcebook for Electronics Calculations, Formulas, and Tables*!

Arranged by subject, information ranges from the simplest elementary operations to the more complex trigonometric and calculus functions. Physical property tables of circuits and materials are included, and many of the formulas are accompanied by application examples to show practical uses. Units conversions, reduced formulas, and "nonconventional" notations are also included to make design work easier and less frustrating.

Professional Reference
440 pages • paperback • 8 3/8 x 10 7/8"
ISBN: 0-7906-1193-7 • Sams 61193
$34.95

Applied Robotics
Edwin Wise

A hands-on introduction to the field of robotics, this book will guide the hobbyist through the issues and challenges of building a working robot. Each chapter builds upon the previous one, extending a core robot project throughout the book. Examples of chapters include: Mechanical Platforms, Power Supplies, Adding Sense, Microcontrollers, Insect Robots, Pneumatics, More Behavior and Intelligence, Programming Projects, Robot Behaviors, and much more.

Electronics Technolgoy
328 pages • paperback • 7-3/8" x 9-1/4"
ISBN 0-7906-1184-8 • Sams 61184
$29.95

To order today or locate your nearest Prompt® Publications distributor at 1-800-428-7267 or www.samswebsite.com

Prices subject to change.

Sams Guide
to Satellite TV Technology
John A. Ross

This book covers all aspects of satellite television technology in a style that breaks "tech-talk" down into easily understood reading. It is intended to assist consumers with the installation, maintenance, and repair of their satellite systems. It also contains sufficient technical content to appeal to technicians as a reference.

Coverage includes C, Ku, and DBS signals. Chapters include How Satellite Television Technology Works, Parts of a Satellite Television Reception System, Installing the Hardware Portion of Your System, Installing the Electronics of Your System, Installing Your DSS, DBS, or Primestar System, Setting Up a Multi-Receiver Installation, Maintaining the System, Repairing Your System, and more.

Guide to HDTV Systems
Conrad Persson

As HDTV is developed, refined, and becomes more available to the masses, technicians will be required to service them. Until now, precious little information has been available on the subject. This book provides a detailed background on what HDTV is, the technical standards involved, how HDTV signals are generated and transmitted, and a generalized description of the circuitry an HDTV set consists of. Some of the topics include the ATSC digital TV standard, receiver characteristics, NTSC/HDTV compatibility, scanning methods, test equipment, servicing considerations, and more.

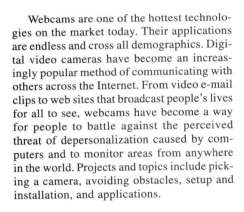

Home Automation Basics
Practical Applications
Using Visual Basic 6
by Tom Leonik, P.E.

This book explores the world of Visual Basic 6 programming with respect to real-world interfacing, animation and control on a beginner to intermediate level.

Home Automation Basics demonstrates how to interface to a home automation system via the serial port on your PC. Using a programmable logic controller (PLC) as a home monitor, this book walks you through the process of developing the home monitor program using Visual Basic programming. After programming is complete the PLC will monitor the following digital inputs: front and rear doorbell pushbuttons, front and rear door open sensors, HVAC systems, water pumps, mail box, as well as temperature controls. The lessons learned in this book will be invaluable for future serial and animations projects!

Electronics Technology
386 pages • paperback • 7-3/8" x 9-1/4"
ISBN 0-7906-1214-3 • Sams 61214
$34.95

Home Automation Basics II:
LiteTouch System
James Van Laarhoven

Daunted by the thought of installing a world-class home automation system? You shouldn't be! LiteTouch Systems has won numerous awards for its home automation systems, which combine flexibility with unlimited options. James Van Laarhoven takes LiteTouch to the next level with this text from Prompt® Publications. Van Laarhoven helps to make the installation, troubleshooting, and maintenance of LiteTouch 2000® a less daunting task for the installer, presenting information, examples, and situations in an easy-to-read format. LiteTouch 2000® is a true computer-control system that reduces unsightly switch banks and bulky high-voltage control wiring to a minimum.

Van Laarhoven's efforts make LiteTouch 2000® a product to be utilized by programmers of varying backgrounds and experience levels. *Home Automation Basics II* should be a part of every electrical and security professional's reference library.

Electronics Technology
304 pages • paperback • 7-3/8" x 9-1/4"
ISBN 0-7906-1226-7 • Sams 61226
$34.95

To order today or locate your nearest Prompt® Publications distributor at 1-800-428-7267 or www.samswebsite.com
Prices subject to change.

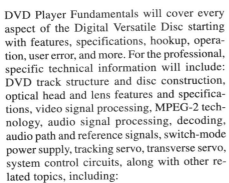

Basic Electricity

by Van Valkenburgh,
Nooger & Neville, Inc.

Considered to be one of the best electricity books on the market, the authors have provided a clear understanding of how electricity is produced, measured, controlled and used. A minimum of mathematics is used for direct explanations of primary cells, magnetism, Ohm's Law, capacitance, transformers, DC generators, and AC motors. Other essential topics covered include conductance, current flow, electromagnetism and meters. Can also be used as a textbook.

Electrical Technology
736 pages • paperback • 6 x 9"
ISBN: 0-7906-1041-8 • Sams 61041
$29.95

Basic Solid-State Electronics

by Van Valkenburgh,
Nooger & Neville, Inc.

Considered to be one of the best books on solid-state electronics on the market, this Revised Edition provides the reader with a progressive understanding of the elements that form various electronic systems. Electronic fundamentals covered in the illustrated, easy-to-understand text include semiconductors, power supplies, audio and video amplifiers, transmitters, receivers, and more.

Electronics Basics
944 pages • paperback • 6 x 9"
ISBN: 0-7906-1042-6 • Sams 61042
$29.95

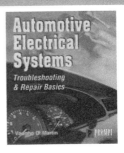

Dictionary of Modern Electronics Technology

Andrew Singmin

New technology overpowers the old everyday. One minute you're working with the quickest and most sophisticated electronic equipment, and the next minute you're working with a museum piece. The words that support your equipment change just as fast.

If you're looking for a dictionary that thoroughly defines the ever-changing and advancing world of electronic terminology, look no further than the Modern Dictionary of Electronics Technology. With up-to-date definitions and explanations, this dictionary sheds insightful light on words and terms used at the forefront of today's integrated circuit industry and surrounding electronic sectors.

Whether you're a device engineer, a specialist working in the semiconductor industry, or simply an electronics enthusiast, this dictionary is a necessary guide for your electronic endeavors.

Electronics Technology
220 pages • paperback • 7 3/8 x 9¼"
ISBN: 0-7906-1164-4 • Sams 61164
$34.95

Automotive Electrical Systems:

Troubleshooting & Repair Basics

Vaughn Martin

Your car has an electrical problem — maybe your entire electrical system is shorting out or your battery is continually drained. Can you fix it yourself? With *Automotive Electrical Systems — Troubleshooting & Repair Basics*, you most likely can!

Automotive Electrical Systems walks you through the reading of repair schematics, which are vastly different on automobiles than on traditional consumer electronics devices. It also helps give you the confidence to buy and change out your car's components. This book stresses electronic test instruments over the more traditional approach to automotive systems, focusing on troubleshooting and repair techniques along with how to interpret your test instrument's results.

Designed for the beginner, it also is a handy guide for the experienced technician.

Automotive
272 pages • paperback • 7-3/8" x 9-1/4"
ISBN 0-7906-1142-2 • Sams 61142
$29.95

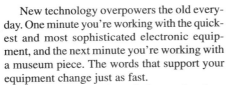

To order today or locate your nearest Prompt® Publications distributor at 1-800-428-7267 or www.samswebsite.com

Prices subject to change.

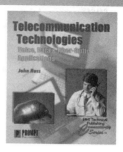

Tube Substitution Handbook
Revised Edition
William Smith & Barry Buchanan

Tube substitution is one of the only feasible methods to repair or restore original tube equipment, but it should not be performed haphazardly. This handbook, itemizing all known vacuum tubes that have been or are still being manufactured, along with their replacements, will make tube substitution not only possible but relatively straightforward and efficient. The most accurate, up-to-date guide available, the *Tube Substitution Handbook* is useful to antique radio buffs, classic car enthusiasts, ham operators, and collectors of vintage ham radio equipment. In addition, marine operators, microwave repair technicians, and TV and radio technicians will find the *Handbook* to be an invaluable reference tool. The *Tube Substitution Handbook* is divided into three sections, each preceded by specific instructions. These sections are vacuum tubes, picture tubes, and tube basing diagrams.

Professional Reference
160 pages • paperback • 6 x 9"
ISBN: 0-7906-1148-1 • Sams: 61148
$19.95

Telecommunications Technologies:
Voice, Data & Fiber-Optic Applications
John Ross

Your job responsibilities are expanding, especially on the jobsite. Those changes in the workplace mean that you need to know more to be effective in your daily tasks. *Telecommunication Technologies: Voice, Data & Fiber Optic Applications* will help you to gain the necessary understanding of this new and emerging technology.
This book contains the information needed to develop a complete understanding of the technologies used within telephony, data, and telecommunications networks. As you progress through this book, you'll learn about the benefits, integration, and future of these technologies. Included are detailed explanations of terminology, comparisons of equipment, viability and cost, as well as applications of the latest transmission media in a variety of business settings.

Communications
368 pages • paperback • 7-3/8" x 9-1/4"
ISBN 0-7906-1225-9 • Sams 61225
$39.95

To order today or locate your nearest Prompt® Publications distributor at 1-800-428-7267 or www.samswebsite.com
Prices subject to change.

Exploring RF Circuits
by Joseph J. Carr

Basic Communications Electronics
Jack Hudson and Jerry Luecke

In this book you will learn about the basics of RF construction, RF circuit design, and gain an understanding of radio frequency circuits. Chapters cover varactor diodes, direct conversion receivers, RF signal generator circuits, RF grounding, RF bridges, and more. With a practical approach and voice, Carr presents up-to-date information and "user-tested" projects that can be utilized by the hobbyist, student or technician.

Basic Communications Electronics discusses how analog electronic devices and circuits are used to create communications systems. Concentrating on semiconductor devices, bipolar and field-effect transistors, and integrated circuits, Basic Communications Electronics will teach you how these devices work and how they are used in analog circuits.

Chapters include: Basic Communications Systems; Analog System Functions; A Refresher; Amplifiers & Oscillators; Modulation; Mixing & Heterodyning; Transmitters; Receiving - Including Detection; Transmission Links; Analog Integrated Circuits; Digital Signal Processing; and more.

Electronics Technology
408 pages • paperback • 7-3/8" x 9-1/4"
ISBN 0-7906-1197-X • Sams 61197
$34.95

Communications Technology
224 pages • paperback • 8-1/2 x 11"
ISBN: 0-7906-1155-4 • Sams 61155
$29.95

To order today or locate your nearest Prompt® Publications distributor at 1-800-428-7267 or www.samswebsite.com
Prices subject to change.

Digital Audio Dictionary
Cool Breeze Systems

Guide to Cabling and Communication Wiring
Louis Columbus

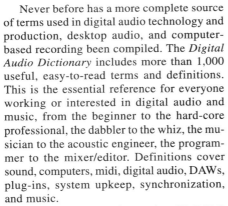

Never before has a more complete source of terms used in digital audio technology and production, desktop audio, and computer-based recording been compiled. The *Digital Audio Dictionary* includes more than 1,000 useful, easy-to-read terms and definitions. This is the essential reference for everyone working or interested in digital audio and music, from the beginner to the hard-core professional, the dabbler to the whiz, the musician to the acoustic engineer, the programmer to the mixer/editor. Definitions cover sound, computers, midi, digital audio, DAWs, plug-ins, system upkeep, synchronization, and music.

Also included is an interactive CD-ROM! The Mac/Windows compatible CD includes an interactive version of the Digital Audio Dictionary, which allows the user to search, print or save found terms for printing later.

Part of Sams Connectivity Series, Guide to Cabing and Communication Wiring takes the reader through all the necessary information for wiring networks and offices for optimal performance. Columbus goes into LANs (Local Area Networks), WANs (Wide Area Networks), wiring standards and planning and design issues to make this an irreplaceble text.
- Features planning and design discussion for network and telecommunications applications.
- Explores data transmission media.
- Covers Packet Framed-based data transmission.

Audio Technology
256 pages • paperback • 7-3/8" x 9-1/4"
ISBN 0-7906-1201-1 • Sams 61201
$29.95

Communications Technology
320 pages • paperback • 7-3/8" x 9-1/4"
ISBN 0-7906-1203-8 • Sams 61203
$39.95

To order today or locate your nearest Prompt® Publications distributor at 1-800-428-7267 or www.samswebsite.com

Prices subject to change.

Designing Power Amplifiers

Stephen Kamichik

This book is the authority on designing power amplifiers! Hobbyists, technicians, and engineers alike will find its contents practical and useful. Designing Power Amplifiers is divided into two sections: Theory and Projects. A detailed circuit description is given for each project. The book also includes problem solutions, a glossary, and a bibliography.

Basic Digital Electronics

Alvis J. Evans

Basic Digital Electronics will teach you the difference between analog and digital systems. The functions required to design digital systems, circuits used to make decisions, code conversions, and data selections are discussed.

Audio Technology
256 pages • paperback • 7-3/8" x 9-1/4"
ISBN 0-7906-1170-8 • Sams 61170
$29.95

Digital Electronics
188 pages • paperback • 8-1/2" x 11"
ISBN 0-7906-1118-X • Sams 61118
$24.95

To order today or locate your nearest Prompt® Publications distributor at 1-800-428-7267 or www.samswebsite.com

Prices subject to change.

Exploring
Microsoft Office XP
John Breeden II and Michael Cheek

Breeden and Cheek provide an insight into the newest and most-advanced product suite yet from Microsoft — Office XP. Office XP is the replacement for Microsoft Office 2000 and is designed to take users into the 21st century. Breeden and Cheek provide tips and tricks for the experienced Office user, to help gain maximum value in this new software.
• Custom installation instructions.
• Full coverage of Excel, Word, PowerPoint, Outlook, Access, and more.
• Covers new features not found in previous versions of Office.

Exploring LANs
for the Small Business
and Home Office
Louis Columbus

Part of Sams Connectivity Series, *Exploring LANs for the Small Business and Home Office* covers everything from the fundamentals of small business and home-based LANs to choosing appropriate cabling systems. Columbus puts his knowledge of computer systems to work, helping entrepreneurs set up a system to fit their needs.
• Includes small business and home-office Local Area Network examples.
• Covers cabling issues.
• Discusses options for specific situations.
• TCP/IP (Transmission Control Protocol/ Internet Protocol) coverage.
• Coverage of protocols and layering.

Computers
408 pages • paperback • 7-3/8" x 9-1/4"
ISBN 0-7906-1233-X • Sams 61233
$39.95

Connectivity
320 pages • Paperback • 7-3/8" x 9-1/4"
ISBN 0-7906-1229-1 • Sams 61229
$39.95 US

To order today or locate your néarest Prompt® Publications distributor at 1-800-428-7267 or www.samswebsite.com

Prices subject to change.